HOW TO SUCCEED
IN ACADEMICS

· · · · · · · · · · ·

HOW TO SUCCEED IN ACADEMICS

· · · · · · · · · · · · · · · ·

Linda L. McCabe
Edward R. B. McCabe

Department of Pediatrics
UCLA School of Medicine
Los Angeles, California

ACADEMIC PRESS

An Imprint of Elsevier

San Diego   San Francisco   New York   Boston   London   Sydney   Tokyo

Academic Press
An Imprint of Elsevier
525 B Street, Suite 1900, San Diego, California 92101-4495, USA
http://www.academicpress.com

Academic Press
An Imprint of Elsevier
84 Theobald's Road, London WC1X 8RR, UK
http://www.academicpress.com

Library of Congress Control Number: 99-65867

ISBN-13: 978-0-12-481833-0
ISBN-10: 0-12-481833-1

PRINTED IN THE UNITED STATES OF AMERICA
07 08 09 10 11   9   8   7   6   5   4

This book is dedicated to our mentors (past, present, and future),
our parents,
and our children.
We thank you for all you have taught us in the past
and look forward to learning even more from you in the future.

Contents

· · · · · · · · · ·

7. PREPARATION OF ABSTRACTS FOR SCIENTIFIC MEETINGS

8. PRESENTATIONS AT SCIENTIFIC MEETINGS: PREPARATION OF EFFECTIVE SLIDES AND POSTERS

9. THE 10-MINUTE TALK

10. THE 1-HOUR TALK, INCLUDING THE JOB APPLICATION SEMINAR

14. MANUSCRIPT REVIEW

15. ETHICAL BEHAVIOR

16. LEADERSHIP

17. PREPARING A CURRICULUM VITAE

18. SUMMARY: GAUGING SUCCESS

PREFACE

Our goal in preparing this material has been to spare you, and others like you, the difficult and sometimes harsh lessons that we learned in our combined academic experience of more than 50 years. Many of the barriers encountered concern highly sensitive areas involving power, control, and gender. These are often not discussed openly. Additional problems may be traditional components of the academic system and may be considered "rites of passage." Other difficulties that are common and systemic may not be recognized as such. When trainees or junior faculty members are confronted with these problems, they may feel that there is something wrong with them personally and that they have encountered, or created, a unique problem that is too embarrassing to discuss even with a trusted mentor. These issues do not occur all of the time, but they make many academicians uncomfortable. Some individuals are reluctant to get involved in seeking a solution. We present examples of situations observed in a number of settings. Unfortunately, such problems are not restricted to the early years of a career in academics. You will find that these issues are not limited to your professional life. Many have features of universality, based on flaws in the human personality. The lessons learned here will apply to your personal life. There are no unique solutions. The most important aspect of an academic career is having mentors for each stage and each aspect of your academic career. In the absence of a mentor who can cover each topic and issue, we hope that this book will help provide guidance. Your attention to these matters will make you a better mentor.

How to Succeed in Academics was originally intended for postdoctoral fellows and junior faculty members, but undergraduates, graduate students, and more senior faculty members have also participated in our course on which this book is based. This book will be useful for anyone contemplating an academic career, whether that person is an undergraduate considering graduate or professional school, a graduate or professional school student, a postdoctoral fellow, or a faculty member.

Academic institutions reward individuals who have publications and bring income into the institution. Too often, superb teachers do not receive appropriate recognition, and the only time mentoring may be considered is in the context of an application for a training grant. We would argue that at each stage of an academic career, an individual needs at least one, and often more than one, mentor. At all times, an individual needs training and guidance in the development of professional skills, whether at a laboratory bench, at a computer, or in a service role. One also needs advice regarding the long-term development of one's career. Advice should be provided by a mentor who is able to transcend his or her own self-interest. Teaching and writing are also important skills for the academician and require mentoring. A trainee or junior faculty member may find one person to fill all of these roles or may need a number of individuals to provide mentoring. Later in one's career, mentoring in developing leadership skills will be worthwhile, but, in fact, leadership is involved among groups of two or three persons working toward a common goal in the laboratory, classroom, or clinic. Most importantly, the academician should develop the ability to provide mentoring to trainees. Perhaps, in addition to celebrating publications and income as key components of success, we should consider the development of students and trainees, and their continued productivity and achievements, in our assessment of faculty members' contributions to their fields.

We began our course in How to Succeed in Academics at Baylor College of Medicine in 1992 to provide group mentoring as part of a training grant for junior faculty members. To groom future participants for the training grant, we decided to include postdoctoral fellows as well. Colleagues, including Ralph Feigin, M.D., Thomas Hansen, M.D., William Shearer, M.D., and Jeffrey Towbin, M.D., participated in these discussions. After two years of presenting this course at Baylor, we moved to UCLA, where we continued to lead these discussions. We also developed a Leadership Workshop aimed at junior to midlevel faculty members. Other discussants involved in the Leadership Workshop were Howard Gatlin, Ph.D., Michael Karpf, M.D., Gerald S. Levey, M.D., Barbara Nelson, Ph.D., Elizabeth Neufeld, Ph.D., Albert Osbourne, Ph.D., Michael Phelps, Ph.D., and Ernest Wright, Ph.D. We presented portions of these programs at professional meetings and at other academic institutions. We learned a great deal from each of these experiences. We also appreciate the input and support of our editor, Craig Panner, in the development of this book.

This work has two purposes: to be a point of discussion for seminars in other institutions and for the individual reader. We learn a great deal from participants every time we discuss these materials. Informal feedback includes the feelings of support participants receive from knowing that others have problems similar to their own and from participating in joint problem-solving. Over the years we have learned a great deal from our colleagues who have served as discussants, and we appreciate their willingness to share their insights.

To assist you in thinking through problems, we provide brief vignettes throughout. We have purposely used gender-neutral names and alternated gender, because these situations occur regardless of gender. These are fictional accounts, although, unfortunately, the occurrence of such problems is all too real. If you are able to gain personal insight while considering these brief scenarios, then they have served their purpose.

Although we focus on problem-solving in academic settings, we recognize that any field of endeavor has its problems. Academics provides the opportunity to focus on an area of inquiry and develop a research program, to share knowledge with students and trainees, to develop professional skills, and to impact a field and society as a whole. Sharing your enthusiasm with trainees through mentoring allows you to have an impact beyond your own efforts, an impact that influences future generations in your field.

Linda L. McCabe
Edward R. B. McCabe

1

• • • • • • • • • •

INTRODUCTORY OVERVIEW

Establishing Personal Goals and
Tracking Your Career

• • • • • • • • • • • • • • • • • •

OVERVIEW

Focus to Achieve Success

Different aspects of your career may progress at different rates. You need to take a realistic, overall view of an academic career and recognize that no one is an immediate expert in all aspects of academic life. Development of expertise in an area may require dedication, one or more mentors, and time. Even with appropriate focus and productivity, however, recognition as an authority in a field requires time that is measured in years, perhaps five to ten.

Do Not Expect Too Much from Yourself
Too Soon

Professional competence in areas such as clinical service and outside consulting is rapid and the rewards are tangible. The danger is that your attention might be distracted by activities that preclude the development of other aspects of your academic career. For example, consulting or seeing patients will provide more immediate gratification, including monetary incentives, than other more academic pursuits. A look back at all the effort and years involved in developing this professional competence will help you to recognize that other aspects of an academic career, such as bench or clinical research, require the same dedication and time to evolve.

Recognition for performing research and writing papers may be more delayed and more personal. Rewards for teaching and curriculum development may be even more nebulous. Committee work and administration can be very time consuming and may go unrecognized. However, each of these activities is important to the development of your academic career and to the maintenance of the infrastructure of your institution and your profession.

Set Goals and Evaluate Them Annually

The realization that you are making progress in different areas of your career may require you to look back one year or even five years. To encourage development of the various aspects of your career, you should set both short- and long-term goals. A long-term goal, such as promotion to associate professor and being tenured, is vague and nonspecific. You stand a better chance of achieving these goals by determining the requirements for promotion and tenure in your institution and planning how to meet these expectations by meeting short-term goals. Review your goals annually: you may be surprised at how much progress you are making toward your longer term goals. If you find that progress is lagging in one or more areas, careful evaluation and review with your mentors will help you develop better strategies for success.

Career Development Takes Time

Kim is a 29-year-old physician beginning a postdoctoral fellowship to receive training in a clinical subspecialty and research. Kim's long-term goal is to be a clinician/researcher/educator in a medical school. Kim is very comfortable taking care of patients. However, entering a research setting for the first time has totally perplexed her. Everyone else in the research setting is very comfortable with their projects, their methods, and their level of competence. Kim's initial research attempts lead to one failure after another. Concerned that research can never be rewarding, Kim meets with her mentor. Her mentor asks Kim to consider how long she spent developing her clinical skills—the four years of medical school, the three years of residency, and the clinical aspects of her current fellowship training. The mentor points out that the other members of the lab group were once neophytes themselves. Kim's mentor suggests that she approach research with the un-

derstanding that there are frustrations in mastering any new skill. With intensive effort on Kim's part and considerable help from other members of the research team, Kim will one day be an effective researcher. Rather than run away from the frustrations of the new research setting to comfortable clinical situations, she needs to put as much effort into research as she once did into learning medicine.

PROMOTION AND TENURE

You Can Not Do It Alone—Identify Mentors

Institutions consider a combination of factors in promotion and tenure decisions. These include national reputation, creative products (publications, inventions), income to the institution (professional fees, grants, donations, patents), teaching, and committee activities. You should have at least one mentor for guidance through the labyrinthine and arcane promotion and tenure process. It would be ideal if this individual was in a different division or department, so that your decisions would not impact their teaching schedule, clinical workload, income, or space. An effective mentor for promotion and tenure should function as a coach, providing plays and pep talks as needed, and always focusing on the goal.

Assignment of Mentors

Dr. Jones is the new chair of a large department with a number of recently appointed assistant professors. Dr. Jones' vision for the department includes the development of a new mentoring program to assist junior faculty through the promotion and tenure process. Dr. Jones assigns mentors who complement their mentees. The mentors can bring their own experiences to the mentoring relationship to assist their mentees in establishing strong foundations for their careers. Examples of pairings will show the complementarity. Sandy is very focused on a new baby. Sandy's mentor is a professor who combined having four children with a successful academic career. Lee really enjoys teaching and receives a great deal of satisfaction from mentoring. Lee's mentor is not only a superb teacher but is also the best-funded researcher in the department. Chris is a very social person, who receives a lot of satisfaction from committee work. Chris' mentor is dedicated to the premises that there is no such thing as a good committee and that new faculty should focus on their teaching and research. Kim is un-

able to complete any task because of a lack of energy and focus. Kim's mentor is a real dynamo, who excels in all areas and is totally unimpressed by Kim's excuses for not fulfilling responsibilities. In addition to these department mentors, each faculty member is also accountable to a research mentor and to Dr. Jones.

Mentors Can Help You Identify Your Strengths and Weaknesses

Each of us must recognize our strengths and weaknesses if we are to be successful. The mentor can assist the mentee in understanding areas that need to be improved by sensitive evaluation and by providing insights into how the mentor developed their own strengths. The mentee should be introspective to help identify relative weaknesses and strengths with the mentor or, in the absence of insightful mentors, to be able to begin the process independently. When no mentors exist locally, the individual may be able to identify potential mentors at the regional and national levels, whose strengths will complement the mentee's weaker attributes.

The mentor should also be able to clarify what a national reputation means and what evidence of national reputation will be considered for promotion and tenure. In any discipline, national reputation would include invitations to speak at other academic institutions throughout the country and at national meetings, publications in nationally recognized journals, review activities and/or editorial board membership for nationally recognized journals, book authorship or editorship, committee membership of national organizations, and/or job offers from other academic institutions.

Your Mentor Must Have Your Best Interests in Mind

Chris has just started as an assistant professor. Chris' mentor is the department chair, who is constantly providing "opportunities" for Chris. These include a heavier teaching load than anyone in the department, several committees that are very time-consuming, and administrative responsibility within the department. Chris is concerned that these other activities detract from his research/creative activities. He worries that when it is time for a promotion and tenure decision, it will not be favorable. When Chris looks at new assistant professors in other departments, he finds that they teach less, participate on only one committee, and have no administrative responsibilities within their departments. When Chris discusses these concerns with the department chair, he is told not to worry. Fortunately, Chris speaks to a senior professor in another department. This professor describes the fate of Chris' predecessors, each of whom has been denied promotion and tenure due to lack of research/creative productivity. Chris and the senior professor meet with his department chair. The three of them develop a plan to decrease Chris' teaching, committee, and administrative responsibilities so that Chris can devote more effort to research/creative activities.

A Mentor Is an Advocate

Sometimes, the more junior faculty member may need to "recruit" a more senior individual to provide mentoring advice and, perhaps, to serve as an advocate. This may be someone with whom the mentee has worked in teaching, research, or committee work, and who the mentee recognizes as a discreet and sensitive person. If you cannot identify such an individual on your own, you may seek assistance from your dean's office, a dean for Academic Affairs, or an ombudsperson.

Timeline for Establishing a National Reputation

Activity	Year 1	Year 2	Year 3	Year 4	Year 5	Year 6
Professional activity	X	X	X	X	X	X
Write articles for journals	X	X	X	X	X	X
Join national organizations	X	X	X			
Apply to make presentations at local and regional meetings	X	X	X			
Apply to make presentations at national meetings	X	X	X	X	X	X
Volunteer for committee membership in national organizations			X	X	X	X
Write review articles for journals				X	X	X
Write chapters					X	X
Write textbooks						X
Apply for positions at other institutions						X

Dual Professional Couples May Have Particular Pressures on Their Careers

The increasing numbers of dual professional couples lead to new pressures on the systems for promotion and tenure. Some institutions are able to focus on an individual's performance when considering promotion and tenure.

Others take advantage of a member of a couple's being "stuck" in the locale due to a significant other and hire, pay, and promote at a lower level than they would if the individual's choices were not restricted. If you are geographically restricted in your mobility, you should assess your institution's treatment of individuals in similar circumstances.

How Do You Determine if Your Institution Is Taking Advantage of You?

Maintain or establish contacts with other individuals at your level within, and outside of, your institution. Professional organizations in your field may publish average salaries. For example, the American Association of Medical Colleges produces an annual salary survey with average salaries for each academic rank, degree, department, region of the country, and type of medical school (public or private). Discuss these issues with colleagues who are five to ten years ahead of you in their careers. Discuss any concerns you have with your mentor.

Institutions May Take Advantage of the Geographically Restricted

Sandy has just finished a postdoctoral fellowship and is looking for a faculty position in the same city. Moving to another city is not possible since Sandy's significant other has a very specialized career and has been fortunate to secure one of the three best jobs in his field in the country. The department where Sandy trained offers her a faculty position as an instructor, nontenure track. Sandy accepts the position since it is the best available alternative. Five years later, Sandy is still an instructor, nontenure track, when a senior member of the department suggests that Sandy apply for a tenure-track assistant professor position available at a nearby institution. Sandy applies for the position, is offered the job, and meets with her current department chair to say she is leaving for a tenure-track position. Sandy's department chair is very upset that Sandy is leaving.

BECOMING AN EFFECTIVE TEACHER

Teaching Is the Core of Academics

The common thread through all of academic life is teaching. While each academician develops their own style, everyone can be an effective teacher if they couple a sincere desire to effectively impart knowledge with sufficient time and effort. The motivating force for teaching is truly caring for and respecting trainees and wanting them to succeed. If you consider the qualities of the outstanding teachers you have had in the past, you can develop your own concept of an ideal teacher with qualities that you would like to incorporate into your own style as an educator.

The Teacher Is a Good Listener, Enthusiastic Encourager, and Effective Motivator

Truly caring for trainees translates into being a good listener, focusing full attention on the trainee. An effective teacher cares about students as people, beyond the training situation. This caring does not imply total acceptance of all qualities of the trainee; rather, the teacher encourages positive behavior and specifies behaviors that should be improved. An effective teacher sets standards that motivate students to do their absolute best. Enthusiastic teachers with a good sense of humor are also effective.

The Teacher Respects the Trainee

Respecting trainees requires a willingness to communicate through explanation or demonstration to clarify information. Effective teachers strive to answer questions effectively without making trainees feel inferior or guilty for asking questions rather than immediately grasping the presentation. Respect for trainees also requires that the teacher is predictable and consistent. The teacher should be willing to say "I don't know" if that is the case. If the trainee presents information suggesting the teacher has been incorrect, the effective teacher demonstrates respect for the trainee and intellectual integrity by admitting that he/she made a mistake.

Knowledge Begets Organization and Confidence

Wanting trainees to succeed requires the teacher to have clear, explicit expectations that are effectively communicated to the trainee. It goes without saying that the teacher should be thoroughly knowledgeable in their area of expertise. A strong knowledge base enables the teacher to be organized and to exude confidence.

The Teacher Is a Learner

You never learn so much as when you teach. Teaching encourages continued growth and professional development. In the course of explaining material you think you have mastered, trainees' questions may be difficult to answer, causing you to present additional explanations or novel examples. You may even have to study and prepare further to deal with the question at a later date.

Be Responsive to Your Critics

While course evaluations may not always be flattering, they are always helpful. If you want to believe the glowing endorsements, you have to accept a grain of truth in the stinging condemnations. Use evaluations to become an even better teacher. You may not agree with each comment, but, if you see a pattern to the criticisms, then you should definitely take heed and make adjustments to your organization and/or delivery. You may discuss this with your mentor and ask the mentor to attend one of your classes to provide feedback.

Seek Evaluations from Your Students

Stacey has had a lot of teaching experience before joining the faculty. Stacey feels that experience is the best teacher and is confident that he is a very good educator. Student evaluations of teaching are optional at Stacey's institution, but he was confident enough to forgo this process. As part of the decision for promotion and tenure, senior faculty solicited trainees' opinions regarding Stacey's teaching abilities. Stacey was regarded by students as ill-prepared, ineffective, and having personal mannerisms that interfered with teaching. Stacey is devastated when confronted by summaries of this information, which jeopardizes his promotion and tenure. If Stacey had solicited student evaluations from the beginning, he may have been able to seek some mentoring for teaching and improve his classroom performance.

Establish Mechanisms for Feedback

In addition to being self-critical, it is extremely important to take advantage of every reasonable opportunity to get meaningful feedback on your teaching. If there is no mechanism or requirement for teaching evaluation in your department or at your institution, then develop a mechanism for yourself. Call on respected colleagues and selected students to give you one-on-one verbal critiques. Follow your students' progress beyond your classes to see if you have given them a firm foundation for future learning.

Course evaluations are like publications and funded grants. They are currency that is irrefutable in promotion and tenure decisions.

Course Evaluations Are Solid Evidence of Your Teaching Ability

Lynn had a lot of teaching experience as a graduate student. Student evaluations of graduate student teaching assistants were required, and Lynn learned much that improved her teaching. As an assistant professor, Lynn continued to use student evaluations, even though they were not required by her department. When Lynn was being reviewed for promotion and tenure, she was asked by the department chair to indicate which full professor in the department should review which aspect of Lynn's performance. There were so few full professors that Lynn had to include Dr. Jones, her nemesis. Lynn suggested that Dr. Jones be responsible for her teaching, figuring that the data from her teaching evaluations were irrefutable. Lynn gave Dr. Jones copies of her teaching evaluations from every semester at the university. Dr. Jones chose to ignore Lynn's teaching evaluations but instead wrote an evaluation based on quotes from students, solicited by Dr. Jones and taken out of context. These quotes were not very favorable and were not in keeping with Lynn's student evaluations. When the department chair read Dr. Jones' letter, the chair summoned and severely reprimanded Dr. Jones. Dr. Jones was required to rewrite the letter, removing the unfair quotes, and substitute a fair summary of Lynn's student evaluations. Lynn's student evaluations were instrumental in her securing promotion and tenure.

Teaching May Not Be the Only Criterion for Promotion

Never assume that superb teaching is "enough" for promotion. It is critical that you know the practical aspects of this process for your academic series (e.g., tenure, research, or clinical track). If research productivity is required in your series, then you must be aware of the operational requirements (e.g., publications and grants).

Just as clinical interactions can provide immediate gratification, so too can teaching for the talented educator. Students respond to and appreciate the effective and enjoyable presentation of information. However, in many institutions, teaching is necessary, but not sufficient, for promotion. You need to know the requirements for your institution and develop a disciplined plan of action for meeting them, utilizing a timeline to achieve intermediate goals. In subsequent chapters, we will provide you with some tips on how to turn your teaching into tangible products that will assist you with promotion.

Professional Development Requires a Teacher to Develop Other Creative Activities

During Lee's first semester of teaching at the university, a senior professor suggested that he request student evaluations so that he could improve his teaching and also have documented proof of teaching effectiveness. Lee developed into an excellent teacher through hard work and changes suggested by student evaluations. Lee really enjoyed teaching and put a lot of effort into preparing lectures and developing new courses. Despite repeated warnings from his department chair that he needed to develop a research program, Lee focused on teaching to the exclusion of other professional activities. When Lee was considered for promotion and tenure, he was turned down due to lack of creative professional activities.

Teach at Every Opportunity—and Plan the Opportunities

You should use every opportunity, both formal and informal, to teach. If someone asks a question and is belittled or ignored, they will not ask a question again and you will lose important opportunities for teaching. Often, questions should be answered briefly with the possibility for more extensive discussion left open. You should probably schedule individual meetings with students that are doing research with you, in order to encourage them to ask questions in a less intimidating setting. These meetings often provide the opportunity for important discussions of topics beyond the research project, such as the students' aspirations and anxieties about their futures.

Learn to Deal Positively with the Disruptive Student

Occasionally, you will have students who ask questions to enhance their self-esteem or to deliberately attempt to embarrass you. Both of these situations are disruptive and need to be dealt with effectively. In either case, you can deflect such questions by suggesting the student review the literature on the topic and make a presentation at the next meeting. Since the motivation of this type of student's questions is not to learn, giving the student more work is usually an effective countermeasure.

The Teacher Is in Control of the Class

Lee was an excellent teacher, always willing to go the extra mile for a student. One of Lee's students was always interrupting Lee's lectures with questions. Initially, these questions were relevant to the lecture and provided the stimulus for some interesting give-and-take discussion between Lee and the students. As the course progressed, the student's questions became more frequent and less focused. Rather than contributing to an understanding of the material, they were peripheral and distracting. The student seemed intent on asking questions that Lee could not answer since they were outside Lee's field. While Lee was willing to prepare answers to questions that he could not answer for the next class meeting, when the questions weren't relevant to the course, Lee wasn't sure the distraction was valid. He discussed this problem with his department chair, who suggested that Lee turn the tables on the student. Every time the student asked a question, Lee asked the student what he thought the answer was. The student didn't know the answer, since many of the questions were unanswerable. When the student couldn't answer, Lee suggested that the student do some research and return to the next class meeting with the answer to the question. By the third time, the student abandoned the questioning. A number of the other students told Lee later that they appreciated his ability to keep the class "on track."

Individuals Can Excel in More Than One Area of Their Professional Lives

We often hear that those who are excellent researchers are not good teachers, but that is simply not the case. There are many examples of individuals who are outstanding educators and investigators. Dr. Louis Ignarro, professor of pharmacology at UCLA, shared the 1998 Nobel Prize in Physiology and Medicine and won the Golden Apple Award ten times for his lectures to second-year medical students.

SELECTING A TRAINING ENVIRONMENT

Choosing a Training Program, Training Institution, and Mentor

OVERVIEW

Choose a Training Program Carefully

How do you decide which training program you wish to pursue? It is worthwhile to spend a lot of time and personal energy in this decision process. If you are fortunate enough to be certain of a particular training program, you should still take the time to consider why you have chosen your career path. This will be one of the main questions on application forms, in interviews with your professors who will serve as references for you, and in interviews with program faculty members and trainees. Simply stating "I've wanted to do this as long as I can remember" rings very hollow. If the training will enable you to meet your goals of developing your talent, helping others, enabling you to better express yourself, expanding and sharing your knowledge base, or providing the tools to solve problems, you have a rational basis for pursuing the training.

Medicine and Science Offer Many
Career Opportunities

Sandy's parents were both physicians. Sandy grew up with everyone asking, "What do you want to be when you grow up?" Her reply was always, "A doctor like Mommy and Daddy." She was always bandaging dolls and pets and running her own "hospital" for them. Sandy took all possible science courses in high school and applied only to colleges that provided a strong premedical education. When Sandy entered medical school, she felt a great deal of pressure to devote all of her attention to medicine. Sandy was conflicted, because she had always enjoyed art as well as science. Whereas previously, there had been time for both school and art, medical school demanded a higher level of commitment. Worried that medicine would cause her to abandon art, she determined that she could combine medicine and art as the writer and illustrator of medical texts.

Questioning Your Decisions Is Normal

You should recognize that you will intermittently question your career path. Any professional degree program is long and arduous, and, at times, your interest may seem to ebb. Don't make any quick decisions to change your direction dramatically at such moments. Also, you should recognize that each field has a wide breadth of opportunity and can accommodate individuals with diverse interests.

Success Requires Passion

To be successful in any endeavor you pursue, you must be passionate about what you are doing. It is difficult to fake passion, but it may be possible to do so during the interview process. The problem will be in the ensuing years if you lack the enthusiasm that is required to be productive and happy.

Questions to Consider in Choosing
a Training Program

You should ask yourself questions such as the following to determine which training program to pursue:

What do you like to do? Do you enjoy teaching? Do you like to work with others or alone? Can you tolerate delayed gratification, or do you want

Schedule for Applying to a Training Program

Years before You Enter the Training Program of Your Choice

	Four	Three	Two	One	¾	½	¼	$\frac{1}{12}$
Begin to consider programs	X	X						
Meet with advisor	X	X	X	X	X	X	X	X
Plan coursework to prepare for training program	X	X	X	X				
Volunteer experience	X	X	X	X				
Research experience	X	X	X	X				
Discussions with people in the field	X	X	X	X				
Search Internet for programs		X	X					
Narrow program selection		X	X	X				
Request application			X	X	X			
Tests for entrance			X	X	X			
Visit for interviews				X	X			
Select program							X	
Move to program							X	X
Begin research							X	X

immediate feedback? Do you want to travel? Do you enjoy writing? Do you want to focus your energy in one area or be involved in a number of areas? Do you want to be at the center of a field or at the interface between fields?

What do you want to be doing professionally 10 years from now? Do you see yourself as a mentor? Are you willing to be in training 10 years from now? When you look at your professors who have completed a similar training program, would you be comfortable in their roles?

What type of setting do you want to be working in 10 years from now? Would you consider leaving academics for the private sector? Are graduates from this training program able to secure positions in settings you would find interesting?

What level of responsibility do you wish to attain? Do you see yourself advancing into an administrative leadership role? What would be your eventual goal?

What level of risk are you willing to assume? How important is financial security? How important is job security? Would the training program mean that you would have to assume a great deal of debt? Would the training program mean a significant delay before you have a "real" job? What is the ratio between number of graduates from the training program and number of openings that you would find interesting? Do graduates move from this training program into jobs or into further training?

Be sure to talk not only with professors but with current trainees, especially first-year trainees, and ask them, does the recruitment for the program match the reality of the trainee program? Ask those close to leaving the program, did the program prepare you for the kind of position you want, are you able to secure such a position, and is there anything that you would do differently if you were to do your training again?

How important is independence in your personal equation? Will the training program provide the flexibility you desire? Will you need further training to reach your goals? Does the training program have different tracks, depending on a trainee's interests?

SELECTING AN INSTITUTION

Use Your Network and the Interview to Help You

The information available on the Internet can empower you in your search for a training program, training institution, and mentor. However, it does not replace the very personal information and insight you can obtain from your advisor, professors, colleagues, and trainees (past and present) in a training program or at an institution. Some programs require an in-person interview. This can be very helpful if you have researched the program and the institution beforehand and go with specific questions regarding the organization of training and the nature of mentoring there. You are judged in your interview by the questions you ask as well as by your answers to questions asked of you. Training programs will evaluate your interest and motivation via your interactions during the interview. While the faculty can answer many of your questions, it is ideal if you can also speak with current trainees to obtain their perspectives. They recently went through the application, interview, and selection process. They can explain why they chose the program they did and discuss whether or not the program is meeting their expectations. They are also very aware of the current prospects for previous trainees or those who have almost completed the program. Be very cautious about any program that does not give you free and open access to its trainees without supervisory faculty present.

Be Sure to Meet with Trainees during Your Interview

Kim applied to a training program and was invited for an interview. Kim's current advisor helped prepare Kim for the interview with a practice session where Kim was encouraged to ask questions about the training pro-

gram dealing with issues important to Kim. During the training program interviews, Kim met with the program director and several senior faculty members. Kim's interviews went very well, with Kim being able to answer as well as ask questions. At the time, Kim thought it was strange that there were no meetings with current trainees, but the positive interactions with the faculty sold the program. After her acceptance and relocation to a new city for the training program, the grim truth became apparent. The professors were totally out of touch with the trainees. While they received a great deal of intellectual stimulation from coursework, their "creative" activities prepared them only to be clones of the professors. Unfortunately, there were 100 graduates of this and similar programs annually, and far fewer academic positions were advertised each year. The unfavorable ratio of graduates to jobs meant that many trainees postponed completion of the program with hopes of becoming more competitive in subsequent years. Some trainees found positions outside academia on their own, since the program faculty were helpful only with academic positions. Other trainees simply left the program after the truth of their situation became apparent. Kim was devastated and immediately began to consider transfer to another institution, even if it meant starting over.

Choose a Program for Its Strengths, Not Its Status

For undergraduates, your choice is between graduate school, professional school, or both. You should ensure that your personal goals are consonant with the goals of the training program. Make sure that your only goal is not to prove something to someone; i.e., choose a program for its strengths, not for its status. The "smart" people select the program that is most appropriate to their personal goals.

Audition the Training Field and the Program

In order to test the validity of your choice of training, you should take electives, perform volunteer or paid work, and enter competitions in the field. These experiences will also provide important letters of reference from those in your chosen discipline. By interacting with others who have completed this type of training program, you will be able to determine:

If the training is appropriate for your goals

If you enjoy interacting with graduates of this type of program—you will spend most of your professional life with your colleagues

If the training program provides the appropriate skills

If the mentors assist in securing the next position

If there are sufficient positions available after training is completed.

Ask Yourself Questions before You Interview

Stacey was high school valedictorian, graduating with a 4.56 average, and was approaching the senior year of college with a 3.96 average. Stacey had always wanted to be a physician. Stacey's parents were both physicians, and both enjoyed taking care of patients in a private-practice setting. Stacey was aware of the rewards and the rigors of medicine and decided that caring for patients was the most gratifying position one could attain. Stacey was a very competitive person and, in applying for medical school, decided to apply to combined M.D./Ph.D. programs because that training was far more rigorous than medical school alone. Stacey had never attempted research outside of high school or college laboratory coursework. With such strong grades, excellent letters of reference from professors, and good scores on the Medical College Admission Tests (MCAT), Stacey was invited to interview by several M.D./Ph.D. programs. During the interviews, Stacey could not answer questions such as: What types of scientific questions excite you? How will you combine subspecialty training with research training? What research experience have you had? What qualities are you looking for in a scientific mentor? How will you feel in four years when the students you entered medical school with are graduating with M.D.s and you have finished two years of medical school and are in your second year of graduate school? Fortunately, several of the interviewers suggested that Stacey amend his application to reflect his interest in clinical medicine and to apply to medical school without the M.D./Ph.D. program.

Choose a Training Program That Will Prepare You beyond Your Training

For those completing graduate or professional school, choice of further training depends on which type of training will support your goals. You will need to ascertain which program requirements are important for success in the academic job market. You need to determine which skills and projects will make you uniquely qualified. Sometimes the most marketable individuals are those at the interface of two fields, and, therefore, you might consider additional training in a different field. You should review the résumés of faculty members in departments that provide the environment you would like. What type of training did they pursue? You need to determine if you will need more than one training situation to prepare you for an academic career. You will need training not only in your area but also in the fundamentals of academic life, such as writing, making presentations at meetings, and teaching.

SELECTING A MENTOR

Choose a Mentor Who Will Mentor

Perhaps the most important choice you make will be the selection of your mentor. Your mentor not only provides training but also assists you in making the next step in your career. While you may have one official mentor, you may need more than one mentor at any one time, and you will need additional mentors for different aspects of your position throughout your career. Among the criteria for selecting your mentor should be the following:

Interest in developing your career

Ability to provide support and training in your chosen field

Modeling of a successful academic career and training in these skills

Commitment to help trainees make the next career move

Success of current and former students in academia

Personal integrity.

A Good Mentor Will Train You for Independence

Too often, trainees select a mentor based upon the mentor's professional standing. While that is important, it cannot replace a sincere interest in, and commitment to, the success of trainees. Success does not simply rub off the mentor onto a trainee; a trainee requires supervision and appropriate experiences in order to succeed. The true mentor delights in the success of his/her mentees. Most of us enjoy having others want to emulate our professional choices, but a true mentor focuses on preparing the trainee for successful independence beyond the current training period. This requires structuring the training experience to encourage the trainee to produce independent products, such as grant proposals, publications, and presentations, that will serve as the currency of their subsequent academic life.

Choose a Mentor Who Is Interested in Your Future

Chris was thrilled to be selected to join the most sought-after group in his training program. The word in the program was that this advisor was on the way to winning the most coveted prize in the discipline. What Chris didn't know was that the prize would come from the desperately hard work of the trainees, none of whom would receive any credit for their efforts. Chris was assigned to do a very repetitive task that was a small part of the very large operation. There was no independent project for Chris to develop or to take with her when she left; there was no opportunity to develop writing, teaching, or other skills. Except for class, all Chris did was the same procedure again and again. Chris' mentor did win the prize while Chris was a trainee. When it was time for Chris to apply for academic positions, Chris' advisor was too busy being celebrated to send letters of reference or speak to colleagues about Chris. In preparing a résumé, Chris found very little to add from this training period. Perhaps Chris' advisor would have won the prize without Chris' help, perhaps not. It is certain that Chris had been used by an advisor who was not a mentor and who had no interest in Chris' career.

Characteristics of the Successful Mentor

How can your mentor help you develop your own independent career?

Encourage excellence

Be sensitive to your needs

Teach principles, judgment, and perspective, in addition to research skills

Introduce you to other colleagues in the field

Identify, and encourage you to accept and adapt to, your strengths and weaknesses

Show you how to recognize and adapt to institutional realities

Provide opportunities for you to develop independence.

For Some Mentors, "Breaking Up Is Hard to Do"

Kim chose a mentor who was very prolific, publishing frequently with trainees and numerous collaborators included as authors. Kim knew that

publishing is key to success in academics and felt that this mentor would provide a jumpstart to his career. Things went along very well, until it came time for Kim to leave the group for a faculty position. It was then that Kim learned that you never really left the group. While you could physically move to a different institution, you were expected to continue your project under your mentor's guidance and to publish with your mentor. Speaking with other former trainees from this group confirmed Kim's impression that moving to a different institution would mean that Kim was simply serving as a satellite for the mentor. Fortunately, Kim spoke to a senior professor at his new institution who suggested that Kim shift to a new project in order to establish independence. The new project involved methods similar to those Kim learned from his previous mentor, but that would enable Kim to establish his independence. Upon learning of this plan, Kim's previous mentor was furious and threatened to ruin his career. In spite of this threat, and with the support of colleagues at his new institution, Kim succeeded in developing independence and became a much better mentor than his own mentor had been.

Consider the Whole Package When Selecting a Mentor

You should never select a mentor based on a single personal characteristic that you have defined ahead of time. Some students will consider only a mentor of one gender. Others prefer a junior faculty member that is very active in the lab, while some trainees prefer the more senior investigator who may have a better network to facilitate finding the next position for the trainee. Looking at only one characteristic of a potential mentor may blind you to other, more important characteristics.

chapter

3

SELECTING A POSITION IN ACADEMIA

Choosing a Department, Institution, and Mentor

NEVER CONSIDER A LATERAL MOVE
EXCEPT

Once you have completed your training, you will be applying for a position in academia. If you have a job, you may be applying for a position at a different institution. Essentially the same criteria apply to both situations. If you already have a position, you should not waste your time and energy applying for a job that would be a lateral move, unless circumstances at your current institution make staying there professionally untenable. Valid reasons for considering a lateral move include power or gender abuse, lack of resources for professional development, inability for promotion, or impossible demands for teaching or service that interfere with professional development.

JOB SEARCHING TAKES A LOT OF ENERGY

You must recognize that looking for a job takes time away from your professional development. Job searches will consume your time, such as when preparing for interviews and travelling, and will require a considerable investment of emotional energy. You should consider moving if it means a promotion, tenure, resources to support your work, a more prestigious institution, better trainees, or the opportunity to improve your career through better interactions with new colleagues. Unfortunately, it may take an offer from another institution to inspire appreciation of you in your current department.

Job Searches Are Ego-Gratifying, but Moves Must Be Carefully Considered

Chris was flattered when Dr. Smith, the chair of a prestigious department at a different university, took an interest in her presentation at a national meeting. After discussing Chris' research, Dr. Smith suggested that Chris apply for an opening for an assistant professor in Dr. Smith's department. Chris was already an assistant professor at a similarly prestigious institution, but the attention from Dr. Smith was flattering. Chris had been frustrated recently when a new assistant professor was hired with a better research package than Chris had received a year earlier. Chris was also concerned that she was carrying a heavier teaching load than the senior professors in her department and that there was not sufficient office staff to process her manuscripts and grant proposals. Chris submitted a curriculum vitae to Dr. Smith and visited his department. Chris did not tell her department chair, Dr. Jones, about her interest in the new job, but simply said Dr. Smith invited her to present a seminar. Chris' interview went well, and Dr. Smith offered Chris the position. Chris was attracted to the position because Dr. Smith offered a large research package, minimal teaching responsibilities, and more secretarial support. When Chris told Dr. Jones she was resigning to take a position with Dr. Smith, Dr. Jones was surprised and upset that Chris would consider another job without first telling him. When Dr. Jones asked Chris why, Chris explained her concerns about her current position. Dr. Jones asked why Chris hadn't mentioned these before and offered Chris more research support (enough to equal the package of the new assistant professor position), a reduced teaching load, and increased access to a secretary. Dr. Jones also suggested that Chris consider the impact on her research program. Shutting down the lab, relocating, and hiring new staff could put Chris six months behind. Considering the career and personal costs of moving, Chris declined Dr. Smith's offer and decided to stay in Dr. Jones' department.

BEWARE OF BECOMING AN ACADEMIC ITINERANT

Even if the institution you are moving to pays moving costs, you still bear the costs of setting up a household in a new location, your work is interrupted, and moving impacts your trainees, employees, colleagues, and family members. The professional disruption can reduce productivity and interfere with your ability to compete for grants. We all know academic itinerants, who move every two or three years without any obvious change in status or reason for moving. One wonders whether they would have been more productive had they not moved so often, and one questions the impetus for all the moves.

DEMAND PROTECTION TO PURSUE THE CAREER PATH AGREED UPON WHEN YOU WERE HIRED

Sam completed training and became a faculty member at the same institution. While Sam had all of the responsibilities of a new faculty member, he still retained the heavy service commitment of a trainee. After a year of frustration and lack of progress, Sam met with his department chair to discuss the situation. The chair agreed that Sam would be hard-pressed to meet the criteria for promotion and tenure if he continued to serve as a trainee in addition to assuming the roles of a faculty member. The chair promised to begin recruiting a new trainee right away and to have one in place in less than a year. Twelve months later, there had been no attempt to recruit a trainee, and Sam was still in an untenable position. He felt a great deal of loyalty to the department and the chair but did not want to fail as a faculty member because of unrealistic expectations. Sam went to the chair and said that he had to begin the search for a position in another institution that would recognize his faculty status and provide trainees to help with the service workload. Sam's job search jolted the chair into reality. The chair immediately divided the service work equally among all of the faculty members in the department and submitted advertisements to the leading professional organizations to recruit not one but two trainees. The chair's willingness to protect Sam's time and the prospect of trainees convinced Sam to forego a job search.

CRITERIA FOR EVALUATING A POSITION

The following are criteria you should use in evaluating a position:

Commitment of the department to the career development of junior faculty members

Sufficient resources to support the professional activities of junior faculty members, providing the basis for their future success and independence

Appropriate expectations regarding time allotted for teaching, administration, and service so the new faculty member has time for academic development

Availability of dedicated mentor(s) to support development of professional skills and to provide understanding of the criteria for promotion and tenure

A larger community with similar interests, both within and beyond the department, to provide the opportunity for exchange of ideas and professional resources and the potential for collaborative efforts

Access to trainees at all levels

Integrity of the department chair—is the offer letter honored?

THE CONTRACT IS ONLY AS GOOD AS THE TWO PEOPLE SIGNING IT

Kim carefully negotiated a wonderful start-up package for his first faculty position. He was offered sufficient funds for him to continue his research, enough space for an office and lab, and a guarantee of 80% protected time for research. When Kim arrived, the funds and space were available as promised. However, instead of spending 20% of his time on service activities, Kim was spending 80%. When Kim spoke to other junior faculty, they all said that the chair had not honored significant portions of their offer letters. When Kim spoke to the chair, the chair said that the letter was written before the unanticipated departure of two senior faculty members and the offer letter was not binding anyway. Kim took his case to the dean. The dean held a meeting with Kim and the chair, where the dean suggested that the chair immediately hire two temporary faculty members and replace the two former faculty members as quickly as possible. The dean stated that once the two temporary faculty members were on board, Kim should spend 20% of his time on service commitments. In addition, the dean would meet with

all other junior faculty members in the department to ensure that the chair
met the promises in their offer letters as well.

THE INTERVIEWEE IS ALSO AN INTERVIEWER

In order to obtain information about a department, you need to ask questions when you visit. You should clearly define the qualities of a department that are important to you and determine whether this department has these qualities, particularly the core features that you feel are absolutely required for your success. You should carefully read all written material the department provides to determine whether the material is consistent and whether the institution shares your values. You should visit the department and the institution Web page, not so much for the "glitz," but for the timeliness of the information and the image of the institution that the Web site provides. You should perform a literature search for publications produced by the department and by the institution, and particularly by any faculty with whom you will interact during your visit. Such information will indicate to the interviewers your sincere interest in the position and will assist you in conversations during formal interviews, as well as during more casual interactions, such as at meals. You should talk to colleagues with connections to the department and the school, since they frequently can give you inside information, but beware of their biases (Why did they leave?) and the timeliness of their data (When did they leave? Have they maintained contacts?). You should also ask colleagues who have no past history with the department or institution their impression of the status, supportiveness, and mentoring environment there.

EVALUATE A PROGRAM'S TRAJECTORY AS WELL AS ITS CURRENT OR HISTORICAL (" . . . IN THEIR OWN MIND") POSITION

Lee received two job offers for her first faculty appointment. One was from a well-established, prestigious private university of long standing. The department had produced a long line of famous investigators over the years. The second position was with a comparatively new state university. This department was growing, and a number of new faculty were recognized as the future of the discipline. When Lee compared the two offers, the private university offered a lower salary, smaller research funding, and heavier teaching load than did the state university. Still, Lee was leaning toward the private university based on prestige and history. Fortunately, Lee com-

pared the NIH rankings for the past 10 years of grant funding. These showed the private university with decreasing money each year and the state university with increasing money. Lee realized if the current trends continued, the state university would have more NIH funding than the private university within two years. When Lee spoke to her mentor, Dr. Smith, about this, Dr. Smith suggested that the dean at the private university was more interested in creating administrative infrastructure than in developing research programs. The dean at the state university had a widely publicized goal of being in the top 10 state universities based on the amount of NIH funding. With Dr. Smith's advice, Lee accepted the position at the state university.

ASK TO MEET WITH A BROAD RANGE OF INDIVIDUALS

Sometimes, you will be asked to suggest individuals with whom you would like to meet on your visit. Because of your research, you should have some ideas regarding colleagues in your area of interest, especially those directing the training program, a relevant center or program, or a core facility. You should also suggest meeting with trainees, recent recruits, the department chair, the division head, and the director of education.

BROAD INPUT WILL ALLOW YOU TO FORMULATE A PLAN

Sandy and Sam were the two finalists for the position of department chair at an outstanding institution. Before their second visit, both were asked whom they would like to meet with on their second visit. Sandy wanted to learn as much as possible about the department and the institution. Sandy asked to meet with scientists that he knew in his area of research, and he also did a literature search to determine additional potential research collaborators. Sandy also queried several friends who knew the department and institution for names of important people in the research and administrative hierarchy. In addition to these people, Sandy also asked to meet with trainees and recent additions to the department. Sam was more concerned about living conditions and salary issues than whom she was meeting. Sam did not ask to meet with anyone. At the end of their visits, the dean asked both Sandy and Sam to formulate a plan for the department. Sandy felt very well prepared to do so, whereas Sam did not know where to begin.

The dean was very impressed with Sandy's understanding of the department and the university, and Sandy was named department chair.

YOUR SEARCH SHOULD BE DATA-DRIVEN

You should also request information regarding the national standing of the department and the institution concerning training, grant or donation dollars, quality of professional activities, individual memberships in prestigious societies, and individual professional awards. In addition to the current data, you should review past levels to determine if the department or school is on the way up or on the way down. Recognize that, particularly for a junior faculty member, the position of the institution and department today is not as important as where it will be in 5 to 10 years, so evaluate trends. Rankings that are most helpful are those done on a per capita basis, which are more revealing in terms of the impact on an individual faculty member. Don't be satisfied with self-serving, unsupported propaganda; ask to see the data.

SUCCESS REQUIRES RESOURCES AND CREATIVITY

Lee was thrilled to be named the division head in a prestigious department in a renowned university. Lee felt that this opportunity would be a springboard for a career that was progressing, but slowly. Once Lee arrived, it became painfully obvious that this administrative position, rather than stimulating his career, would probably end it. As division head, Lee was responsible for a training program for which there were no funds from the department and for which there was no staff support. The only way to support such a program was through service income, including individual honoraria. Before taking the position, it had never occurred to Lee to question why none of the current faculty wanted to become division head. Now Lee understood. But Lee also had experience with a little-used federal funding program and knew of a niche group of foundations in his discipline. Lee was able to pitch proposals to this funding network and developed a new resource base for the division.

Selecting Grant Opportunities

Understanding the Organization of the NIH, Other Governmental Entities, and Private Foundations

Overview

Cultivate Your Network of Program Officers

When you begin the search for grant support, your best allies are the agency representatives. They can inform you of upcoming "requests for proposals" (RFPs) or "requests for applications" (RFAs) and advise you of the type of proposal or application most relevant to your area and the stage of your career. If you applied for a grant and were unsuccessful, then they may be able to help you understand a critical review of your proposal and why you did not receive support. With this insight, you can restructure your proposal and submit a responsive, revised proposal.

If you have not yet submitted an application, then the agency representatives will be able to describe the type of feedback you can expect to receive from the review process. Private foundations may simply inform you of your success in securing funding. In those cases, you may not receive feedback that you could use to improve the likelihood of success for a future effort.

Be Sure There Is Funding for a New Initiative

With any announcement of a grant, it is worthwhile to call and speak with the representative indicated in the announcement to ensure they are interested in receiving your proposal and that they have money. Even with a positive response to your questions, there is no guarantee that you will be funded. You can also ask how many proposals have been received in the past and how many proposals will be funded in the current competition.

The assurance of the availability of funds for a new grant program is particularly important. There have been times in the past when new initiatives were announced and proposals accepted, but no proposals were funded because there was no money for the program. There never was any intention of funding proposals for the initiative: the announcement was simply a "trial balloon" to demonstrate interest in the topic to key individuals in the agency.

Meet Short Deadlines—Do Not Procrastinate

If a program announcement is made which is relevant to your area and your stage of career, you should certainly apply if your agency contact feels it would be appropriate. If a new program is announced and you determine that funding is available for the program, you should apply, even if there is little time until the deadline. The first time an announcement is made, there may be less competition due to the inability of some groups to mobilize and submit a proposal in time. Your contact in the agency may make you aware of upcoming program announcements before they appear. This will give you a running start on your proposal.

How Do You Find Out about Program Announcements?

One source of information is the Internet. Web sites include information about the agency or foundation, representatives and how to contact them, funding priorities, types of support, blank proposals, currently funded proposals, and program announcements. The following Web sites may be helpful:

National Institutes of Health (NIH): http://www.nih.gov

NIH Guide (published weekly—includes requests for proposals and policy information): http://www.nih.gov/grants/guide/index.html

Maternal and Child Health Bureau: http://www.hrsa.dhhs.gov/grants. htm

Agency for Health Care Policy and Research: http://www.ahcpr.gov

Centers for Disease Control and Prevention: http://www.cdc.gov/

March of Dimes: http://www.modimes.org/

Robert Wood Johnson Foundation: http://www.rwjf.org/grant/jgrant.htm.

In addition to searching the Internet for program announcements, you can also find them in professional publications. Other suggestions for funding sources can come from discussions with senior faculty members in your area or a review of their funding portfolios. Funding sources are often acknowledged in publications or presentations, so careful review of the Acknowledgements may give insight into new funding opportunities. Some institutional grants and contracts offices include facilities or personnel to help with your search for grant support. They may even forward announcements of grant opportunities via e-mail or newsletters.

ORGANIZATION OF THE NIH

Develop Your Familiarity with the NIH

We are most familiar with the NIH since 90% of the support for research in departments of pediatrics comes from the NIH. The NIH consists of two arms, Program and Review. These were designed to be distinct and separate, in order to disconnect decisions about the programmatic needs of the NIH from decisions about scientific merit of, and funding for, individual applications. Program includes the various Institutes that determine funding priorities and issue requests for proposals. Review includes the Center for Scientific Review (CSR; formerly, Division of Research Grants) and the review groups, the Study Sections. Descriptions of Institutes and Study Sections can be found at the NIH Web sites, http://www.nih.gov/icd and http://www.csr.nih.gov/refrev.htm, respectively.

You should establish relationships with the relevant program officers in the appropriate Institutes at the NIH, i.e., those with priorities that include your area of research. They can advise you regarding the relevance of particular program announcements or certain types of proposals. Each Institute will have specific RFPs/RFAs, unique funding mechanisms, and particular interpretations of training and career development awards. Institute program officers will be able to assist you in selecting the relevant study section for review of your proposal. Your mentor(s) and senior faculty in your research area can also provide insight into Study Section selection. If you are responding to an RFP or RFA, the review group will already be determined, and review may be carried out within the Institute and not by CSR.

NIH Program Officers Are Valuable Resources

Sam was an M.D. postdoctoral fellow with plans for a career in academic medicine. Sam's mentor suggested that Sam apply for an NIH National Research Service Award (NRSA) to receive NIH support during training and to begin to establish a grant portfolio with the NIH. Sam's mentor had previous M.D. postdoctoral fellows who had been successful in obtaining NRSAs from the National Heart, Lung and Blood Institute (NHLBI) and from the National Institute of Child Health and Human Development (NICHD). Because of Sam's research interest, she applied to the National Institute of Diabetes, Digestive and Kidney Diseases (NIDDK). Sam's application was rejected by NIDDK because they reserved NRSAs for postdoctoral fellows with extensive laboratory experience. Sam had performed some clinical research but had never worked in a basic science laboratory. Sam's mentor's previous fellows who were successful in obtaining NRSAs from NHLBI and NICHD had similar backgrounds. When Sam spoke to the program officer, she learned that NIDDK liked Sam's proposal. However, NIDDK would not fund Sam's research at the level of an NRSA. Instead, they asked Sam to revise the proposal for submission as a Mentored Scientist Clinician Development Award (MSCDA), a grant to provide support for advanced postdoctoral fellows that would continue into their early faculty years. NIDDK actually had two types of MSCDAs, one for individuals like Sam with no or limited bench experience, and another for those with extensive basic science research training. Sam did apply for an MSCDA and was successful. If Sam or Sam's mentor had discussed Sam's NRSA application with the NIDDK program officer before submission, they would have saved all the effort expended on the NRSA application, submitted an MSCDA application instead, and received funding sooner.

Choose the Best, Not the Easiest, Study Section for Your Proposal

In selection of a review group for your proposal, you should review the descriptions of Study Sections that seem appropriate. Does the description of the Study Section mesh with your research plan? Are you familiar with the research of the members of the Study Section? When you review the publications of the members of the Study Section, do you find several individuals who publish in similar areas and who ask questions similar to yours? If they are interested in a similar area of research, then this would be a Study Section that might be appropriate for review, but you must be sure to cite their work if it is appropriate. Don't be frightened off by "tough" Study Sections. Experience has shown that they typically provide fair, constructive reviews, while "easy" Study Sections may be more arbitrary, less consistent, and less informative. If you find a Study Section that is a good match with your research, you should indicate your choice in the cover letter you submit with your proposal. Present the rationale for your choice using keywords found in the description of the role of the Study Section. You are in a better position to understand your research program than is someone performing a cursory review and triage in the CSR for the tractor-trailerloads of proposals received at each grant's submission deadline.

Constructive Criticism from a "Tough" Study Section Will Lead to Eventual Funding (if You Pay Attention)

Lee had a funded NIH research grant. Study Section A had reviewed Lee's grant, and was considered to be tough but fair in their review. Lee had had to submit two amended applications before Study Section A scored his grant in the fundable range. Lee was preparing an application to NIH for a second research grant. This application was on a different topic but with essentially the same methodology as his funded grant. Lee could request Study Section A, but a senior colleague, Dr. Smith, suggested that Study Section B was easier. Study Section B had given Dr. Smith a score in the fundable range on Dr. Smith's initial application that was funded. The members of Study Section B were all long-time colleagues of Dr. Smith. Since Dr. Smith was one of Lee's collaborators, Dr. Smith indicated that this would improve Lee's standing with Study Section B. Lee decided to request Study Section B. Unfortunately, members of Study Section B did not consider Lee to be the person who should be undertaking the research he proposed. They gave no constructive criticism of Lee's proposal, and it did not receive a priority score. Lee was stunned by what he felt was the capriciousness of Study Section B. Without any help from Study Section B, Lee did not know how to amend his proposal to receive an improved score. Lee requested Study Section A for his first amended application. While Study Section A did not give Lee a fundable priority score, they did provide excellent feedback. Lee revised the proposal in accordance with the recommendations of

> *Study Section A and resubmitted it to them. This time Lee's research was funded. Lee decided that he would stick with Study Section A for future applications.*

You May Request Exclusion of a Reviewer

What do you do if the Study Section you've selected includes a bitter professional rival? When you receive notification of your Study Section assignment, write a letter to request that that individual not be allowed to review your proposal and not be in the room when the Study Section discusses your proposal. This can be done even if the person is the chair of the Study Section. Do not be so naive as to assume that the person would rise above your past disputes and give an objective evaluation or that they would openly declare the conflict and disqualify themselves from the review. This may be just the opportunity they may have been looking for to discredit your research program.

Do Not Trust Your Nemeses to Recuse Themselves

> *Chris and Kim were colleagues in the same department. Although their topics of research were different, they did apply similar methods in their research and preferred the same Study Section for review of their grant proposals. By chance, they submitted research grants at the same time to this Study Section. When reviewing the list of Study Section members, Chris noted that Stacey was a member. Both Chris and Kim had previous collaborative efforts with Stacey, and each of them had ended their collaboration after Stacey took major credit for work that had been an equal effort by both parties. Chris decided that when the appropriate Study Section was assigned to her grant proposal, she would ask that Stacey not be allowed to review it or to be in the room when the rest of the Study Section discussed it. Chris informed Kim of Stacey's membership on the Study Section and her plan to prevent Stacey's review of her proposal. Kim laughed, and told Chris that she was overreacting and that Stacey would declare the previous collaboration conflict and disqualify himself as a reviewer. Kim felt that even Stacey would not be so unprofessional as to purposefully give a bad review to a former collaborator. Both Chris and Kim were assigned to Stacey's Study Section. Chris wrote asking that Stacey have nothing to do with the review of her grant, but Kim did not. Chris received a reasonable*

review with a priority score within the fundable range. Kim's review was very negative, very personal, and not at all constructive. Kim's score was so far out of the funding range that there was little hope of ever being funded. Chris received support for her proposal; Kim had to revise and resubmit. On the second submission, when Kim was again assigned to Stacey's Study Section, Kim requested that Stacey not be allowed to review her grant and be out of the room when the grant was discussed. This time, Kim received a positive review, a priority score within the fundable range, and funding. While Kim would never know who reviewed her first submission, Kim did know that Stacey did not review her second submission. Kim also noted that Chris was funded nine months before she was. She resolved that if any future proposals were directed to a Study Section with Stacey as a member, Kim would immediately request that Stacey not be allowed to review her proposal.

Request up to Three Institutes to Improve Your Chances for Successful Funding

In addition to specifying a Study Section in the letter of transmittal with your proposal, you should also specify up to three Institutes that would be interested in funding your proposal. While the program officer you have been talking to represents only one Institute, they may be able to suggest other Institutes with similar funding priorities. Your mentor and senior research colleagues will also have helpful suggestions. The publications and presentations of others in your research area will often specify the Institute supporting their research.

Institutes vary in the amount of funds they have and in their ability to fund specific areas. Funding priorities of the Institutes are listed on the Internet (http://www.nih.gov/icd). If you name more than one Institute and your first choice does not have the money to support your proposal, another Institute may have more resources and be able to step in and provide funding. The program officer will help you to complete the paperwork necessary for multiple Institute assignment.

Multiple Institute Assignments Pay Off

Sam was submitting a research grant to the NIH. Sam knew that Institute A was the main Institute supporting other researchers in the department and indicated Institute A as his first choice. However, Institute A was relatively impoverished compared to Institute B, which also covered the topic area of Sam's application, so Sam added Institute B as a second choice of Institute. Sam received a reasonable funding percentile, the 17th percentile, on his proposal. Unfortunately, the percentile was two points above the funding range of Institute A, which was up to the 15th percentile. However, Institute B was able to fund proposals up to the 20th percentile, so Sam's proposal was funded by Institute B. If Sam had not included Institute B in his submission letter and had not requested the multiple Institute assignment, his grant would not have been funded.

Occasionally Applications Are Lost, So Protect Yourself

In addition to sending the requisite number of proposal copies indicated in the instructions, you should send a courtesy copy to the program officer who helped you with your application. Include a cover letter indicating the Study Section and Institutes you specified in your proposal cover letter. Keeping the program officer informed is important, but it also serves as a check on the NIH grant assignments. You should receive notification of assignment to a Study Section and Institute within four to six weeks after the proposal due date. If you do not receive such notification, you should call your program officer and ask them to check on your proposal. If you are assigned to a different Study Section or Institute than you requested, you should immediately write the CSR stating your case.

Supplemental Letters Update the Reviewers on Your Progress

If the Study Section and Institute are the ones you requested, you should contact the Scientific Review Administrator (SRA) named in your notification to determine the deadline for submission of additional materials to the Study Section and limitations on the length of such a communication. Since investigators are awarded NIH grants for what they have accomplished, you need to show evidence of your productivity and commitment to the project by updating your Study Section with new data and publications compiled since your original submission.

Courtesy Copies of Grants Pay Off

Sandy applied for a research grant to ease the transition from fellowship to faculty. During the application process, Sandy received excellent advice

Timeline for Establishing a Research Career

(Years are numbered; Fel–Fellow; Fac–Faculty; PI–principal investigator)

Activity	Fel 1	Fel 2	Fel 3	Fac 1	Fac 2	Fac 3	Fac 4	Fac 5	Fac 6	Fac 7
Select mentor	X									
Apply for training grants	X	X								
Do research	X	X	X	X	X	X	X	X	X	X
Publish	X	X	X	X	X	X	X	X	X	X
Apply for grants to transition from fellow to faculty		X	X							
Apply for faculty positions			X							
Identify a mentor for transition to faculty			X	X						
Establish affiliations with centers, institutes, and training programs				X	X	X	X	X	X	X
Apply for funding as an independent investigator				X	X	X	X	X	X	X
Apply for funding as part of a program project						X	X	X	X	X
Apply for funding as PI of a program project, center, or training program									X	X

from a program officer at NIH. When Sandy submitted his proposal, his mentor suggested that a courtesy copy be sent to the program officer who had been so helpful. A month later, the program officer called Sandy to ask when the grant would be submitted, saying that the copy she had received looked very promising, but they had never received the final submission via the CSR. Somehow, the proposal had been lost in the NIH system. Sandy had a receipt from an overnight courier service indicating submission to the CSR on the same date the courtesy copy had been sent to the program officer. Fortunately, the program officer was able to have the grant considered in that review cycle and the grant was funded. Without the courtesy copy sent to the program officer, Sandy would not have known that his grant had been misplaced, and the grant may not have been considered during that cycle.

Long-Term Goal: Establish a Research Career

The following table shows how to begin, as a first-year fellow, to establish a research career. While this is meant as a general guideline, note that at all times you must have the time for research and must publish the results of that research.

5

· · · · · · · · · ·

WRITING A GRANT

Selecting the Specific Aims, Preparing the Budget, and
Developing the Research Proposal

· · · · · · · · · · · · · · · · · ·

OVERVIEW

You Will Never Get a Grant for Which You Do Not Apply

We have enjoyed some success in obtaining grant support for research and training. However, our success rate is probably no higher than most. We follow the same instructions as those coined for voting in a certain city: we just apply "early and often." Once when we were celebrating a grant award, a colleague suggested that we must be getting every grant for which we applied. We invited him to see our file drawers full of rejected grant applications. There is no guarantee that you will be awarded every grant you apply for; but one thing is certain—you will never receive a grant if you do not apply for it.

Success Requires Confidence, Focus, and Persistence

Important factors for success in obtaining grants are confidence, focus, and persistence. You need to be confident enough in your ideas and methodology to submit a proposal, and if you are initially unsuccessful, to respond to criticism and submit a revised proposal. Your proposal should focus on a single area, one in which you have experience and the preliminary data and publications necessary to support an application. You need to demonstrate your persistence by revising and resubmitting your proposal for the next cycle, with new data and publications.

Do Not Let a Bad Review Shatter Your Confidence

Stacey was always very successful in school and in the top 10% of every training program. Because he enjoyed the research required as part of his training and found teaching very rewarding, Stacey decided to pursue an academic career. During his last training program, and with the help of his mentor, Stacey submitted a grant to support his research. Stacey had enjoyed several successes with previous grant proposals for earlier training support and for funding to attend professional meetings, so when the review came back on Stacey's most recent proposal, he was devastated. The review began with a scathing condemnation of his mentor and proceeded to destroy point by point Stacey's preliminary results and research plan. In addition to the absence of any constructive criticism, there were a number of glaring errors of fact in the review. Stacey's mentor suggested that the review was a personal attack on him, not Stacey, Stacey's preliminary data, or the proposed research. The mentor argued that since there were many factual errors, Stacey's request for a re-review would be honored by the NIH. Despite the support of his mentor, Stacey did not request a re-review and left research altogether. Stacey allowed a vindictive review committee member to rob him of the joy he felt in doing research.

PREPARING A GRANT REQUIRES AN ORGANIZED PLAN

The following questions must be addressed as you plan and prepare any grant proposal.

Who?
What?
When?
Where?
How?

Considering these issues formally and effectively early in the process will clarify plans for your application.

Who?

This question is not trivial. The principal investigator (PI), the person responsible for writing the proposal and carrying out the proposed research once there is funding, needs to be established initially. We have all observed instances in which the person writing a grant was not listed as the PI. For purposes of promotion and tenure, a junior faculty member needs to have independent grant funding and an independent research program. The junior faculty member should be the PI for most of the grants he/she writes. Early in your research career, as a fellow or faculty member, you may prepare a grant providing salary for yourself that will be submitted listing your mentor as the principal investigator (PI). This is acceptable as long as there is a very specific plan to develop an independent research career for the fellow or faculty member, and the mentor gives credit for the junior member's contribution in letters of recommendation. For experience, a trainee, not the mentor, should write their own proposals for training support. It is generally quite easy for reviewers to distinguish an application prepared by a trainee from one written by a more seasoned investigator. Part of the training experience is preparing grant applications so that the trainee will be able to secure independent funding before the end of the training period.

Finally, you should be very careful regarding the designation co-principal investigator (Co-PI). The NIH recognizes only a single PI and the designation of Co-PI, since it can reflect a wide variety of relationships, is generally considered to be meaningless. We have all observed pairs of supposedly equal collaborators where one is always the PI and the other is always the Co-PI. Only the PI receives credit in the department, and in the institution, for the grant. Truly equal collaborators should alternate PI status on their grants so they will get equal credit for promotion and tenure, as well as laboratory and office space considerations. If you find yourself in a collaborative relationship with another investigator, particularly if that individual is senior to you, and your attempts to alternate PI status are rebuffed, you should consider walking away from the collaboration since it is quite likely that you are being used. When promotion, tenure, and space decisions are made, your colleague will receive the majority of the credit for the work and benefits based on your efforts. We have observed numerous examples of such relationships, in which the junior partner, who was consistently designated the Co-PI, was not promoted, did not receive tenure, and was not provided with the space required for research. If you find yourself in such a situation, you should consult your mentors to help you to evaluate objectively whether you are being used inappropriately and whether to terminate the collaboration.

Choose Carefully When Selecting Your Collaborators

Chris and Kim were excited to be hired by a very prestigious department that was attempting to develop their area of research. As beginning assistant professors, they found themselves sharing a large laboratory and decided to pool their efforts in order to move their research forward. Rather than each of them identifying their own mentors, they decided to mentor each other. They were careful to divide the research according to their different technical expertise. They felt that they would have a better chance of obtaining grant support for their research if they submitted a proposal presenting their combined expertise and preliminary data. They flipped a coin and Chris won the toss, so he was the PI and Kim the Co-PI. The understanding was that Kim would be the PI on their next proposal. Their original proposal with Chris as the PI was funded. Suddenly, the equal collaboration began to change. Chris wanted more control of laboratory personnel and dominated lab group meetings. It soon became clear that a single grant would not support their joint research efforts. Kim suggested that they apply for a second grant with her as the PI. Chris went ballistic and angrily accused Kim of threatening their collaboration. Kim reminded Chris that they had agreed that Kim would be the PI on a second grant. Chris claimed that he had been carrying most of the weight of the research and deserved to be PI on the second proposal. When Kim went to the department chair and asked for assistance in mediating the disagreement, the department chair agreed that Chris should be the PI. The department chair also indicated a concern regarding Kim's research productivity. Finding that other senior faculty in the department and the dean shared the view of the department chair, Kim found a position in another institution and changed research projects. In the new position, Kim identified a research mentor to assist with establishing her new research program and to advise her regarding future collaborations.

In addition to establishing the PI, potential collaborators need to be identified. These are individuals who have specific skills in methodology or who direct core facilities providing services needed for the proposal. You can not be an expert with demonstrated published skill in every method. You will need to rely on collaborators to teach you certain methods or to provide certain services in order to accomplish your proposed research. You will need a letter agreeing to be a collaborator from each person you identify for this role.

Modern Science Requires Collaborators—and the NIH Requires Their Documentation

Sandy was applying for her first R01. Since the R01 is an independent research grant, Sandy mistakenly concluded that she needed to do all the work herself. She prepared a very careful and thorough proposal based on her previous work in the area. The proposal involved a large variety of methods, most of which Sandy had not yet attempted. She was confident that she could master these new methods. However, these methods required a great deal of equipment that Sandy did not have, so she included requests for the equipment in her proposal. The budget for equipment was so large that it exceeded the total budget for most R01s. Sandy barely finished the application in time to meet the submission deadline. She had not asked her mentor, Dr. Smith, or other senior faculty members at her institution to review the proposal before sending it in. After the proposal had been submitted, Sandy did give a copy of the proposal to Dr. Smith, who gave the proposal a thorough review. Dr. Smith asked to meet with Sandy to review her proposal. He knew there was a problem when he looked at the total amount requested for the first year on the first page of the proposal. Review of the proposed budget confirmed Dr. Smith's initial concern. When Dr. Smith read the proposed studies, he knew the budget reflected Sandy's insistence on performing all of the techniques herself. Dr. Smith suggested to Sandy that "independence" did not mean working in a vacuum, but, rather, referred to research design, especially creativity in crafting and testing hypotheses. Dr. Smith emphasized that Sandy's research would be seriously delayed by the investment of time required to order all of the equipment, set it up, and establish the methods in her lab. Dr. Smith informed Sandy that there was a core facility on campus that provided method A in a very cost-effective fashion with high quality output. Method B could be accomplished in the laboratory of a colleague at another institution who had established the method after 10 years of development. Method C had been developed by Dr. Jones in their department, and Dr. Jones could teach Sandy this method. Dr. Smith explained that Sandy should have listed these scientists as collaborators, obtained letters of collaboration from each of them, and included the letters in the grant proposal. Sandy's grant was not scored, and the Study Section echoed Dr. Smith's recommendations. By following these suggestions, Sandy obtained a fundable score from the Study Section with the next submission and her amended application was funded.

If you are requesting support for personnel in addition to yourself on the grant, you need to demonstrate their expertise through their NIH biographical sketch and in the budget justification section of the application. Professional colleagues should be unpaid collaborators whenever possible.

What?

In answering this question, you first need to decide on your research topic and determine whether the topic is appropriate to the request for proposal and the mission of the granting agency to which you are applying. You then need to determine the specific aims of your proposal. These are very specific, focused research objectives that can be accomplished collectively within the time frame of the grant funding period. The specific aims are determined by your hypotheses that are to be tested. All NIH grant proposals are hypothesis-driven. The NIH has been much less interested in supporting technological breakthroughs except as they support hypothesis-driven research or particular mandates, like the Human Genome Project. The specific aims detail what the proposed research is intended to accomplish. Each specific aim should target a key scientific question. There should be between three and five specific aims for a 5-year proposal. You should use your specific aims to organize the rest of your proposal, since this will provide a consistent framework to assist the reviewers in following the structure, logic, and flow of your application.

When?

You should use the following table as a guide to answering this question.

Timeline for Preparing and Submitting a Grant Proposal over a 5–Month Period

Activity	Month 1	Month 2	Month 3	Month 4	Month 5
Obtain application and read instructions	X				
Discuss interest in proposal with program officer	X				
Literature review	X	X	X	X	X
Develop Specific Aims	X	X			
Discuss Specific Aims with mentor and colleagues	X	X	X		

Activity	Month 1	Month 2	Month 3	Month 4	Month 5
Perform pre- liminary experiments	X	X	X	X	X
Write literature review	X	X			
Write pre- liminary experiments		X	X	X	X
Write pro- posed experiments			X	X	X
Develop budget				X	
Collect biographical sketches and other support			X		
Request letters of collaboration			X		
Review proposal and budget with mentor and colleagues				X	
Make changes recommended by mentor and colleagues					X
Draft letter of transmittal in- dicating type of grant, NIH Institute(s), and Study Section					X
Reread instruc- tions and assemble application					X
Send courtesy copy to program officer					X

Send update of progress on research program when appropriate

You may not always have five months to prepare a grant application, but your initial application will always take longer than you expect. Some investigators like to leave everything to the last minute. You can not do this and expect to meet your deadline with a complete, thoughtful, well-organized, and neatly prepared application. Preparation of a grant proposal is a test of your ability to plan and organize in order to accomplish the research you are proposing and an opportunity to demonstrate these positive traits to the reviewers. You must bring as much care to preparing the proposal as you will bring to carrying out the research. If you leave everything to the last minute, you will not be able to present a compelling argument for your future success. You are vulnerable to system failures, such as the crashing of the departmental computer network, jamming of the copier, lack of toner for the printer, or illness of the person responsible for preparing the grant. All of these problems have happened to us, in addition to many others. Grant agencies rarely are sympathetic to your requests for an extension, except in cases of serious illness. Requests for extensions based on poor planning will give the agency and the reviewers an initial negative impression about your ability to organize your research in a productive manner. They have many more proposals submitted than they can fund. If you take yourself out of the running by not submitting your proposal on time, that makes a negative decision on your grant that much easier. Grant agencies subscribe to the notion that "procrastination on your part does not constitute an emergency on their part." Although you may have heard of investigators who hand carry their proposal on an airplane to the grant agency to meet the deadline, this is not an advisable alternative.

Where?

The question of where to submit your application is already established if you are responding to an RFP that specifies the review group and Institute. If you are not responding to an RFP, then suggest the Study Section and Institutes in your letter of transmittal included with the proposal. You should also talk with your mentors about other agencies that would be potential funding sources for the same, or a similar, proposal.

How?

The issue of how the research will be accomplished needs to be demonstrated by the organization of the application. Neatness definitely counts. Grant reviewers tend to be mature professionals who are donating their time. They appreciate a clear application among the multitude of applications they review. If you can effectively educate the reviewers about your area of proposed investigation, they will be impressed.

An important indication of your ability to carry out the proposed research is your timeline for accomplishing your specific aims. The reviewers will evaluate how realistic your time frame is. Most of us are overly ambitious and allow too little time for successful accomplishment of each specific aim and subaim. It is rare for the reviewers to criticize an application for proposing too little research for the grant funding period.

PREPARING THE GRANT PROPOSAL

The Specific Aims Represent the Organizational Structure for Your Grant Application

The grant proposal should be based on, and organized around, your specific aims. These aims need to be hypothesis-driven and testable. They should be based on a new idea and should be brief and clear. They serve as the organizing outline for each of the other sections of your proposal. Remember, the person reviewing your grant is reviewing many others. Anything you can do to improve the clarity of your proposal will enhance the reviewer's ability to follow your thought process and understand your planned research.

The Specific Aims Should Be Focused but Independent

One of the major criticisms of junior investigators is that they propose a lifetime of work in a two-year proposal. The specific aims should be reasonably accomplished within the time frame of the grant proposal. Another major criticism is that the specific aims are structured so that if the first specific aim is not achieved, then no work can be done on the other specific aims. The aims need to be interrelated and focused on achieving a common goal and testing a single overarching hypothesis, but they need to be independent, so that the second and subsequent specific aims can be carried out even if the first is not achieved.

Specific Aims That Are Sequentially Interdependent Will Be Fatally Flawed in Review

Kim prepared a grant proposal with three specific aims. The first involved the development of a knockout mouse with a particular phenotype that would mimic a human disease. The second specific aim provided developmental characterization of the affected mice. The third aim attempted gene therapy for the knockout mouse. Kim's mentor, Dr. Jones, reviewed the proposal and noted the dependence of specific aims two and three on specific aim one: if the mouse could not be made or if it had no phenotype, then the aims could not be achieved. Kim stubbornly refused to rewrite the proposal and submitted the one Dr. Jones reviewed. The Study Section did not score Kim's proposal, citing the impossibility of accomplishing any of the proposed research if Kim could not generate the knockout mouse with the human phenotype. They went on to list a number of human disorders for which there was no appropriate mouse model, in spite of multiple attempts by many different research groups.

Emphasize the Significance of, and Rationale for, Your Proposal with a Balanced Literature Review

The background and significance section is not simply a review of the literature, but, more importantly, provides the rationale for your proposal. The rationale will be more compelling if it states the logical foundation for each specific aim. Rather than leaving it to chance whether the reviewer will recognize the argument for each specific aim, make it clear by organizing the background and significance section according to the specific aims. Your goal is to convince reviewers that your proposed research is important and that one outcome of your research will be to advance knowledge in the field. You want to clearly demonstrate the novelty and potential significance of your proposed work. As part of your literature review, determine whether any members of the Study Section you are requesting have published in the area of your research. Their relevant publications should be cited in your proposal. If you don't cite them, they will assume that you do not think their work is important. This could affect their review of your proposal. Similarly, provide a balanced review and cite the publications of others equitably, or the reviewers may be concerned about your ability to be objective in your research.

Read the Primary Literature and Go Back More Than 5 Years

Lee was in a hurry to write and submit a grant proposal with a deadline of October 1, 1999. Lee only had time to do an online search of the literature published within the past 5 years. Rather than obtain the papers to read herself, she used the abstracts downloaded as part of the literature search. Lee did not ask anyone at her institution to review her proposal before submission. The reviewers assigned to Lee's grant were experts in this area. They noted the superficiality of her literature review and that Lee had not cited any articles published before 1994, in spite of the fact that this had been an active area of investigation since at least 1980. They chastised Lee for proposing experiments that had already been performed, published be-

tween 1990 and 1993. They pointed out that Lee did not correctly interpret a number of the cited studies. Apparently, the abstracts had not provided Lee with enough background to correctly evaluate the research. They were concerned that she did not address some of the basic controversies in the area that had been presented in review articles before 1990. The reviewers did not score Lee's application and Lee was very embarrassed. Lee should have undertaken a more thorough review of the literature, had her mentor and senior colleagues review the proposal, and submitted the proposal for the next deadline. Submitting a proposal before it is ready does not save time; it usually puts you behind schedule. It also does not enhance your reputation. When these reviewers see Lee's name in other contexts, such as reviewing her abstracts for presentation at scientific meetings or her manuscripts for journals, they will remember Lee's very superficial literature review and wonder if she is superficial in the performance and evaluation of the research as well.

Preliminary Results Demonstrate Ability and Feasibility

In the preliminary results section, you should demonstrate the feasibility of your specific aims through your own research. Like all other sections, it is best to organize your preliminary results according to your specific aims. This section also showcases your relevant expertise, particularly with state-of-the-art techniques. You need to show that you are the person to perform this research and that you are already doing so. The data should be unequivocal, should be presented clearly, and should include a critical discussion of any of its limitations. It is important to have a substantial volume of preliminary, unpublished data showing your momentum in this research area. However, if you have so much data that you could have one or more publications from the material in the preliminary results section, you may be criticized for not being able to generate publications from your work. Figures, graphs, and tables are helpful to support your scientific credibility. When you are applying for a $500,000 grant, you should include color figures in the proposal if they are the best presentations of your data.

Preliminary Data Are Critical for Successful Review

Chris and Lynn are each preparing a grant proposal. Chris includes detailed descriptions of the methods in the proposed research section and doesn't have room for figures and tables summarizing his data from preliminary research. Lynn incorporates data from preliminary studies that illustrate her ability to perform these methods, so she just mentions the standard methods she has already demonstrated she can accomplish. The reviewers fault Chris for too much experimental detail and not enough preliminary results. Lynn is congratulated by the reviewers for her preliminary data. Lynn is funded, Chris is not.

The Research Design and Methodology Section Should Not Be a Cookbook, but an Opportunity to Demonstrate Your Logical Approach

Your Research Design and Methodology section should be organized according to your specific aims and should be presented in enough detail to permit reviewers to evaluate what you will do and why. There needs to be a rationale for each specific aim. You also need to discuss what you will do if your hypotheses are not confirmed. Do not detail standard methods that have already been published. It is best if you can demonstrate that you or your collaborators have published experiments using the methods and technologies proposed. Reviewers are always concerned whether the investigator will know how to process data and design future experiments. In order to be sure that you remember to discuss the logical progression of your data, you may wish to include separate sections entitled "Anticipated Results" and "Alternative Approaches" at the end of the presentation of each experimental design for each specific aim or subaim.

WRITING STYLE

Neatness Counts

In writing your proposal, you want to be very clear. One reason to have colleagues review your proposal is to help you find grammatical errors that may obscure your science. You should spell check your proposal because misspellings will be considered a sign of sloppiness that may carry over from your writing to your research.

Try to Teach, Not "Snow," Your Reviewers

Your proposal needs to be written in straightforward language, without jargon. You need to realize that while the reviewers may be in your general field of interest, their expertise is probably not in your very specific area.

Avoid Pages That Are "Walls of Text"

The appearance of your proposal is important. A page full of text with no spacing is daunting. You should break up the text with spacing, headings, schematic diagrams, and, best of all, data (figures or tables). The NIH specifies a minimum size of type font and the maximum number of characters per inch. The NIH will check the type font and will return a proposal to you if your proposal does not meet their criteria. Note that the characters per inch specification is taken as an average, and the minimum font size they allow (10 pt.) may be too small if the font is a "proportional" one with different spacing for each letter.

Follow the Rules Carefully

One proposal Lee submitted was actually returned. It used the minimum size type font that the NIH instructions specified. However, the letters were of different widths ("t" took up less space than "w"). This is called proportional type. Lee checked portions of the text to determine whether he met the criterion for number of letters per inch. The sections he checked were fine, so Lee submitted the proposal. However, someone at the NIH scanned other sections of text where there must have been many words like "individual" and not so many words like "mandatory." Rather than call and alert Lee to the problem, the NIH mailed the proposal back with instructions to change the font or the proposal would not be sent to the members of the Study Section for review. When Lee called the NIH to determine the deadline for the reformatted proposal, Lee was told that the proposal had to be back at the NIH the next day. Faced with that deadline, Lee did not have time to revise and reduce the length of the proposal but was only able to switch the typeface to one that met the minimum size requirement and was not proportional. Unfortunately, to stay under the maximum number of pages, Lee had to eliminate all spacing between paragraphs. All of his hard work to present an attractive grant proposal was canceled by the switch to a different font. Lee's grant received a very favorable score that was less than one point above the funding cutoff. Lee wondered if switching to the less attractive font and eliminating any spacing between paragraphs had impacted his score enough to push him out of the fundable range. Should Lee have argued with the NIH when his proposal was returned? Definitely not, since decisions by the agency cannot be appealed. Regardless of how arbitrary their rules seem, you agree to follow the rules when you submit your proposal.

BUDGET AND BUDGET JUSTIFICATION

The Budget Needs to Be Realistic

The budget represents another opportunity to demonstrate that you will be able to carry out the proposed research. There needs to be sufficient time and effort planned to complete the project. Each item in the budget requires adequate justification. You should adhere to the total and specific limits in the grant instructions. If the budget is specified, you cannot change it. If there is no budget specified by the agency, then you should consider the guidelines that we will discuss. Reviewers will cut from your budget items that exceed the prescribed, or commonly recognized, upper limits or items that are excluded by the funding agency. If you exceed prescribed limits, then the reviewers will assume that you cannot follow directions. If limits are prescribed but one or more areas of your budget are excessive, they will assume that you are naive or that you think the reviewers are naive. Aim for a total amount within the range of other funded grants of the same type held by your colleagues.

New Modular Budgets for the NIH

The NIH has just begun a modular budget program for a number of different types of grants. With this system, you specify the personnel and their percent effort as well as pieces of equipment, their model numbers, and cost. You do not have to list salaries or itemize supplies for the budget pages. You can use the system that follows to determine the cost to perform the research you are proposing. As part of the modular system, you will specify your annual budget in increments of $25,000 up to a maximum of $250,000. The budget justification describes your personnel and equipment and explains any difference in total direct cost from one year to the next.

The Personnel Budget Needs to Assure Reviewers of Success

The personnel budget should include salary and fringe benefits, which can often be 30% or more of the salary. You may contact the department payroll person to determine the appropriate fringe benefit rates for personnel you wish to include. You should include salary for yourself, if it is allowed, and you will, in fact, be able to devote that level of effort. Certain grants specify the percent effort you need to commit. Discuss this level with your department chair to ensure that you will be able to fulfill the requirements of the grant if your proposal is successful. It is unwise to include your percent effort at too low a level, or it will be assumed that you will not be able to devote sufficient time and energy for the success of the project. Generally, the effort should be 20% or greater for an R01 or similar project. You should also recognize the option of putting a lower salary on the grant than your percent effort, but again you should discuss this with your chair to be sure it is acceptable. Because you are applying for your first grant, you will not be able to conceptualize the possibility, but senior investigators with multiple grants need to be careful not to have more than 100% effort total on all of their grants. The technical personnel you list should be appropriate to the project. If you do not have a technician for this project, you should name a technician like the one you would hire if funded. Since you need extensive preliminary data to get a research grant, it is likely that you will have someone in your group who helped you acquire these data and who has the appropriate experience. It is easy to cut a "To Be Announced" item from a budget, but more difficult to cut a person with demonstrated skill in areas relevant to the research. Administrative personnel are limited to NIH program projects and center grants and are not permitted on individual research grants. Consultant costs are generally discouraged unless you have a compelling justification for them and can communicate this well. It would be better to make the person a collaborator at no salary or, if their contribution will be substantial, include him/her as a coinvestigator.

The Equipment In Your Budget Should Be Used Solely for This Project

Any equipment requested should be specific to the project. Reviewers feel that the institution should provide laboratory set-up equipment, so they will rarely fund equipment such as freezers and refrigerators. You should provide the name, source, and model number of the item needed, as well as the price. NIH reviewers generally expect that equipment will be purchased during the first year. Any exception to this needs to be carefully justified. One exception that may be successfully argued is the instance in which a method is to be introduced in a later year, but this must be explained—you cannot assume that the reviewers will understand this from reading the body of your grant.

Carefully Itemize Your Supplies

Supplies should be itemized by category: glassware, animals, chemicals, radioactivity, etc. Animals should be itemized including the species, age, purchase price, and cost of care (daily rate and number of days). In general, you are allowed between $10,000 and $15,000 per full-time equivalent (FTE) in personnel. To determine how many FTEs you have, total the percent effort from the budget page. Budgets increased beyond these limits may be possible when there is extensive animal or tissue culture utilization, for example, though these need to be carefully and thoroughly justified.

Travel Should Be Limited and Focused

Only domestic travel for the principal investigator is permitted at the rate of $1000 to $1500 per trip. Travel should be to a meeting you are attending to present the results of your research. National meetings are preferred over regional meetings. Patient-care costs are generally frowned upon, as are alterations and renovations and consortium/contract costs.

Other Expenses Should Be Carefully Considered but Not Inflated

Other expenses can include publication costs (figure preparation, journal page charges, reprints), warranties for equipment, radioactive and toxic waste disposal, and service contracts. Only centers and program projects can budget for telephone, postage, express mail, and duplicating.

Any Drastic Changes in Budget from Year to Year Will Attract Reviewers' Attention and Concern

The budget for the entire project period should include no drastic changes after the initial year. There should be no major equipment after the first year, unless there is a very specific justification for it. The NIH does allow for a 4% increase in all categories except equipment each year.

The Budget Justification Is a Chance for You to Demonstrate Ability and Insight

Budget justification should be specific and should include sufficient explanation. The role of each person should be detailed, as well as their previous training and experience relevant to the project. The use and importance of each piece of equipment need to be described. The use of each category of supplies and other expenses should be detailed. Justification of travel to a scientific meeting should indicate the meeting and the purpose of your attendance, preferably to present the results of your research.

Minority/Disability Supplements Represent "New Money" to Encourage the Careers for Appropriate Individuals

As part of a commitment to encourage individuals who are members of underrepresented minority groups or who are disabled to enter research careers, NIH allows investigators with two or more years of support remaining on their research grant to apply for a Minority/Disability Supplement. Minority/Disability Supplements fund the salary and some supplies for trainees who are members of such groups. These funds are in addition to funds approved for the grant and provide for an additional individual to work on the project, preserving the research dollars for other personnel, equipment, and supplies.

Trainee Funding Is an Important Research Support Mechanism and Will Help Establish You as a Mentor

Equally important to a junior investigator are applications for trainee support. These can be made directly to the NIH by an individual trainee, such as the National Research Service Award for postdoctoral fellows. Other trainee support goes to a department or institution to support a training program, and the trainee applies to the training program on campus.

Many Center Grants Have Funding for Junior Investigators

Some center grants (for example, the Cancer Centers and Mental Retardation Research Centers) include small grants for junior faculty members as well as access to core resources. Not only is this funding and access helpful financially, but the additional mentoring and opportunity to interact with senior faculty in the center is invaluable. There also may be a "halo effect" operating when a center awardee applies for their own grant.

c h a p t e r

6

• • • • • • • • • •

GRANT REVIEW

How Review Groups Work, Responding to the Reviewers'
Feedback, and Preparing the Revised Application

• • • • • • • • • • • • • • • • • •

OVERVIEW

If Your Grant Was Assigned to the Wrong Study Section, Contact the Scientific Review Administrator (SRA) Immediately

The notification of assignment to a Study Section will include the name, address, and telephone number of the scientific review administrator (SRA) assigned to the Study Section. If the Study Section to which you were assigned was not the one you requested, you should immediately call the SRA of the Study Section to which you were assigned and the SRA of your preferred Study Section. You should then write a letter requesting a change to the other Study Section. In this letter, include the key words from the description of your preferred Study Section and note how your proposal is concerned with these topics. This letter should be addressed to the center for scientific review (CSR), with copies to the SRA of both Study Sections.

If Your Grant Was Assigned to the Correct Study Section, Contact the SRA about the Format and Deadline for a Supplemental Update

If you have been assigned the Study Section and Institute(s) that you requested in your letter of transmittal (discussed in Chapter 4), you should immediately call the SRA to determine the format and deadline for supplemental materials for your proposal. Since the SRAs are responsible for sending copies of the update for your proposal, they determine the length and the due date of your update material. You will be awarded an NIH grant based on what you have accomplished in the preliminary data of your proposal and in your update. If you are truly committed to the research program you have outlined in your proposal, you will have made progress between the time you submit your proposal and the SRA's deadline for supplemental material. In the absence of such progress, the reviewers will question your commitment and your ability to carry out the proposed research. You might argue that you should not be expected to have made progress in the absence of funding, but, in general, reviewers are not sensitive to this argument. In your letter you need to summarize these new data, update the status of publications listed as submitted or in press in your proposal, and include copies of any manuscripts submitted since your proposal was sent. Remember, this is an update, not a rehash of your proposal. You will usually have about four months between notification of your Study Section assignment and the due date for your supplemental materials.

A Supplemental Update Will Demonstrate Your Ability to Progress

Lee and Stacey are new assistant professors in the same department. They both submitted a grant proposal with the same deadline. Lee's mentor encouraged Lee to submit a proposal update. Lee continued to work very hard in the lab in order to have new data for the supplemental material letter. Lee also revised a manuscript that was tentatively accepted after the original proposal was submitted. Fortunately, a letter of acceptance of the revised manuscript arrived before Lee sent in the supplemental materials. Stacey's mentor did not mention a proposal update, and, because this was Stacey's first proposal, she did not know about the possibility of submitting more material. Even though Stacey accumulated more data, since no update letter was filed with the SRA, the Study Section was unaware of Stacey's progress. Lee, however, had a great deal of data to show, as well as the copy of the in press manuscript that had been at the submitted stage when the proposal was sent. The primary and secondary reviewers of Lee's proposal were very impressed with Lee's progress. Lee scored at the 15th percentile, while Stacey earned only a 40th percentile rating.

Timeline for Preparation of a Proposal Update

Activity	Month 1	Month 2	Month 3	Month 4
Continue research	X	X	X	X
Prepare manuscripts	X	X	X	X
Submit manuscripts	X	X	X	X
Revise manuscripts	X	X	X	X
Prepare data for update			X	X
Write update				X

How Study Sections Work

A Study Section Is Made Up of Regular and Ad Hoc Reviewers, and, Occasionally, Consultants

Your proposal will be assigned to a primary reviewer and a secondary reviewer. Usually, these are members of the Study Section, but they may be ad hoc reviewers, individuals with appropriate expertise who may be under consideration for the Study Section and who will attend its meeting. Occasionally there is not the appropriate expertise to review your grant among the members or ad hoc reviewers in your Study Section. In this case, one or more consultants will be asked to review your proposal and prepare a commentary but will not be present at the meeting of the Study Section.

Not Everyone in the Study Section Will Review Your Application in Detail

At the Study Section meeting, the primary reviewer will spend two to five minutes presenting their opinion of your proposal. Usually, the secondary reviewer will only present points of disagreement with the primary reviewer. While the primary and secondary reviewers are presenting their opinions, the rest of the Study Section may leaf through your proposal. Members of the Study Section who are not the primary or secondary reviewers of your proposal typically only read the abstract, though some may be more thorough. They may casually glance through your proposal, looking for specifics during the presentations of the primary and secondary reviewers. The nature of the review by the majority of the members of your Study Section reinforces the importance of the abstract. It needs to be clear and must include your specific aims and the significance of your proposed research. In addition, your proposal needs to be well organized and your budget justification should include sufficient detail so those members of the Study Section who are not the primary or secondary reviewers will be able to understand your proposed budget. Organization of each portion of the grant, with headings and subheadings for each specific aim and subaim, will facilitate understanding by the members of the review group during the presentation of the primary and secondary reviewers. The members will also look for tables and figures to examine preliminary data.

The Primary and Secondary Reviewers Generally Will Set the Range of Your Score

After the presentations of the primary and secondary reviewers, your proposal will be scored. The range of scores is between 100 and 500. As in golf, the lower your score, the better. The primary and secondary reviewers will give their scores first, with other members of the Study Section typically following their lead. Occasionally, another member of the Study Section will disagree with the scoring of the primary and secondary reviewers. When the scores of the primary and secondary reviewers or another member of the Study Section differ substantially, there will be a broader and longer discussion to try to achieve consensus, although this is not always reached, and then the chair will instruct the membership to "vote your conscience." The scores of all members of the Study Section will be averaged.

Your Feedback Will Come in Written Form and May Be Supplemented by Discussion with Your Program Officer

You will receive your score and the comments of the primary and secondary reviewers approximately six weeks after the meeting of the Study Section. The program officer from your primary Institute will also be at the Study Section meeting so he or she may be able to help you interpret the comments of the Study Section more quickly and in more detail than is provided in the written comments. Since the reviewers' comments are written prior to the review and are frequently not revised to reflect the discussion, the written critiques may not be reflective of the score. A brief section at the beginning of the review is intended to provide this information, but the information depends on the skill and conscientiousness of the SRA.

Your Percentile Rank Is More Important Than Your Raw Score

Some Study Sections are more rigorous than others. The scores from each Study Section are rank ordered and combined with those from other Study Sections for calculation of a single percentile ranking for all of the proposals from all of the Study Sections. This is intended to correct the score inflation of some Study Sections. This final percentile is communicated to you and the Institute(s) for your proposal. Some Institutes have more money than others and are able to fund more proposals, and therefore they can fund proposals of a higher percentile. If the first Institute you requested is not able to fund proposals up to the percentile for your proposal, hopefully the second or third Institute you requested will have more money, will fund to a higher percentile, and will be able to support your research.

The Discretionary Zone (DZ) Gives an Opportunity for the Institute's Advisory Council to Shape the Institute's Program

Some Institutes have a mechanism called the discretionary zone (DZ). For example, an Institute might have the ability to fund all grants up to the 23rd percentile, but may choose to fund automatically to the 20th percentile, and have all grants in the 20th to 30th percentiles placed into the DZ. Members of the Council, which advises the director of the Institute, will review all grants in the DZ and rank order them for funding priority. These decisions will not necessarily retain the original order by percentile ranking established by the Study Sections but will consider programmatic issues (e.g., areas in which more research is indicated). Differences in rigor and advocacy may exist between different Study Sections (e.g., when one Study Section gives a better written impression of the application but a much poorer score that may not be normalized by the percentile score). The Council can adjust for these discrepancies.

If Your Proposal Is Triaged, You Should Seek Guidance in Deciding How to Proceed

Half of all proposals will not be scored by the Study Section; they will be "triaged." You will receive the primary and secondary reviewers' comments, but you will not know where your proposal ranked. Were you at the 51st percentile or the 99th percentile? Since your proposal was not discussed, your program officer will be unable to provide any additional information beyond the primary and secondary reviewers' comments, but from their experience they may be able to help you interpret the comments. Since you are able to submit only two amended applications of a proposal, the interpretation of the comments by your program officer will prove invaluable, and your mentors will help you to know how to respond. The goal of the triage system is to permit more discussion of the better grants proposals, but it places a significant proportion of applicants at considerable disadvantage.

Do Not Give Up on Yourself

Chris and Lynn each submitted a grant for the same review cycle. Both grants were triaged, devastating both Chris and Lynn. They spent a lot of time consoling each other, realizing that they had only two more chances to resubmit these proposals. Chris resubmitted at the next opportunity, while Lynn was still licking her wounds, unable to focus on research or grant writing. Although Chris did not receive funding for the first amended proposal, Chris was not triaged and received a ranking of the 30th percentile. Encouraged by this, Chris filed a second amended proposal, which was funded. Lynn still hasn't revised her proposal and has decreased her research effort.

CRITERIA FOR GRANT REVIEW

The members of the Study Section are charged with reviewing proposals in accord with the following criteria. Before submitting your proposal, you should consider each of these five points and include them in your proposal.

I. Significance

While your research project is important to you, it may not be significant in the larger context of the NIH. Too often junior investigators are so focused on their day-to-day research that they cannot stand back to determine how their research fits into the discipline. While you may be pursuing a particular line of research in order to satisfy your scientific curiosity, this is not sufficient justification for the NIH to support your research. If you feel this is important basic research that will become even more important based on your contributions, you will have to argue your position aggressively and effectively in your application. You need to show that your research will advance your field.

II. Approach

Your research design and methods need to provide the answers to the testable hypotheses derived from your specific aims. Your research design needs to include consideration of viable alternative hypotheses and alternative experiments and methodologies. Your methods should be standard, proven techniques, or, if you are developing new methods, you need to include demonstration of their reliability and validity.

III. Innovation

Your approach to your research should be novel. After a thorough literature search, you should propose specific aims that demonstrate your unique approach to this area of research. You should employ new methods and technologies. You may even challenge the status quo in the field if you have sufficient rationale and preliminary data.

IV. Investigator

Your qualifications to pursue this research are demonstrated throughout the grant. Your biosketch includes your training, experience, and publication record. You should develop realistic budgets and timelines to encourage the reviewers' confidence in your ability to accomplish your specific aims. A reasonable budget and duration are also critical. The Study Section reviewers are also researchers. Their own laboratory experience and their review of numerous grants provide them with standard expectations for proposal budgets. They readily spot any deviation from their expectations. If you do have an unusual expense in your budget, you need to completely justify its inclusion. Include a timeline at the end of your proposal. This demonstrates that you can plan and that you have the experience to recognize how long it will take to accomplish each portion of your specific aims.

Your Budgets and Timelines Demonstrate Your Ability to Plan Realistically

Lee and Lynn are second-year assistant professors, and each is applying for their first NIH research grant (R01). They are performing similar research techniques that should have similar budgets. However, they have two very different approaches to developing budgets for grant proposals. Lee's mentor suggested an annual budget of $150,000 as the average for an R01 in their field. Lee thought she knew better. Having seen one funded application from someone in a different field for $300,000 per year, Lee inflated the costs of her research twofold. In addition, Lee decided to request $200,000 in equipment for year 1, which included everything on Lee's long-term equipment "wish list." Even though Lee had access to most of this equipment in the core facilities of the department, she wanted her own. Lee's proposal was not funded. The reviewers stated that she showed an inability to plan for her proposed research and that her budget was significantly overinflated with equipment that would be underutilized for this project and should be available in the form of a core resource as an institutional commitment to this area of research.

Lynn took a very different approach to budgeting, not wanting to ask for too much money. Lynn's budget was about $100,000/year. Unfortunately, Lynn was not very careful in documenting mouse costs. Lynn failed to include the new increase in daily mouse care, the new university rule limiting investigators to no more than four adult mice per cage, and the fact that he would want to separate adult mice by sex until it was time to breed. Lynn did not research the costs for the core lab facilities to derive each transgenic mouse and knockout mouse. Lynn's proposal was approved and funded at the level Lynn requested. However, Lynn soon realized that he needed to apply for grants from private foundations in order to be able to fund his research and accomplish his specific aims.

Your preliminary data show that you have already partially accomplished your specific aims. Your choice of methods, anticipated results, and alternative approaches indicates that you can engage in critical problem-solving to ensure completion of the grant. Your inclusion of collaborators suggests that you are able to utilize experts in the pursuit of your research.

V. Environment

The resources available for your use need to be carefully documented. On the resources page, you need to detail the equipment that you can use already available in your laboratory and core facilities. If you are applying for a training grant, the scientific credibility and past trainee history of your mentor are critical. You should demonstrate a viable community of scientists ready to assist you with your research through your collaborators. You also need to show evidence of institutional support through core equipment and the start-up funds you used to equip your laboratory.

The Mentor's Skill in Mentoring Must Be "Sold" to the Reviewers

Dr. Jones suggested that his postdoctoral fellow, Sam, prepare an application for an NIH individual National Research Service Award (NRSA). Part of the NRSA application was a listing of Dr. Jones' previous pre- and postdoctoral trainees. When Dr. Jones completed this section, he was very strict in interpreting the guidelines and listed only those trainees who were in Ph.D. training programs or who had completed M.D.s or Ph.D.s and were in formal postdoctoral training programs. Sam's NRSA application was not funded, and one of the criticisms was that Dr. Jones did not have sufficient mentoring experience. For Sam's amended application, Dr. Jones included less traditional trainees, such as Master's degree students, medical students performing summer research projects, M.D. postdoctoral fellows spending six months on a research project with Dr. Jones, visiting scientists coming for three to six months to do research in Dr. Jones' lab, and an associate professor from another institution taking a one-year sabbatical with Dr. Jones. Sam's amended NRSA proposal was funded, and the reviewers had high praise for Dr. Jones' mentoring abilities. The primary difference between the original and the amended applications was that Dr. Jones carefully documented all previous mentoring experience

RESPONDING TO THE REVIEWERS' COMMENTS

If Unsuccessful with This Submission, Pay Attention to the Reviewers' Comments, Since They Will Guide Your Resubmission

When you submit a grant proposal, you have put in a great deal of work over a long period of time on a research program that is very important to you. If your proposal is not funded, it is normal to feel rejected, angry, confused, depressed, disappointed, and just plain rotten. As you read through the comments of the reviewers, you may feel that they could not have read your proposal and say what they are saying. It is easy to dismiss their comments as totally erroneous but, in every review, there is some truth or at least a misperception that you did not dispel. You should put the comments in a drawer for a day or so until you have calmed down and can read them more dispassionately. The worst thing you can do is to call the SRA of your Study Section and rant and rave about the unfair review your application received.

The second worse thing is to complain to the program officer of your Institute.

Be Positive in Your Interactions with Your Program Officer

Kim worked for years to cultivate a relationship with the program officer of the Institute most likely to fund Kim's research. The program officer helped Kim decide which type of grant to apply for and answered Kim's questions regarding the proposal. The program officer suggested that Kim apply for another grant in response to a request for application. When the review of Kim's first grant was not positive, Kim immediately called the program officer to complain and suggested that the problems with Kim's proposal were due to following the program officer's advice. Even though Kim called the next day to apologize, the relationship that Kim had with the program officer was never the same. The program officer was disappointed with Kim's behavior. When the program officer was selecting a special review panel in Kim's area of research, the program officer could not recommend Kim given such unprofessional conduct.

Use Your Network to Help You Craft Your Revision

When you have calmed down enough to engage in a rational discussion with your mentor, you should do so. You may also wish to seek the advice of other senior scientists in your area. These individuals will be able to advise you regarding the validity of the comments and recommend how you should respond to the review. You need to determine how to address the criticisms in order to improve the reviewers' perceptions of your work and how to craft the revision of your application to "sell" it to them more effectively.

Your Mentors May Be More Objective in Their Evaluation of Your Reviews

Sandy was totally devastated by the review of her proposal. Sandy's initial reading of the review suggested that she did not know how to formulate testable hypotheses, do a literature review, conduct research, or present a research plan. Sandy was so embarrassed that she would not have shown Dr. Smith the review, but Dr. Smith asked whether Sandy had received the review. Dr. Smith's interpretation of the review was totally different from Sandy's. Dr. Smith pointed out that the reviewers suggested only the restructuring of Sandy's specific aims, the addition of one citation to her literature review, a possible second interpretation of one of Sandy's preliminary results, and two additional methods under the second specific aim. Dr. Smith noted that each of these suggestions was constructive, and advised that Sandy incorporate them into the amended application and include her additional preliminary data. Dr. Smith urged Sandy to submit an amended application for the next deadline. The same information was there for Sandy and Dr. Smith to read in the review. Sandy took the criticism too personally, while Dr. Smith's experience with grant proposals enabled him to be more objective in the interpretation of the reviewers' comments.

Contact Your Program Officer if You and Your Mentors Are Concerned about Serious Flaws in Your Review

Occasionally, flawed reviews can occur when one or more reviewers allow personal feelings to guide their review. If you find a large number of factual errors in your review and the tone of the review is malicious, you should discuss a possible appeal of your review with your mentors. Be sure that the review is flawed; a difference of opinion is not sufficient basis for an appeal. You do not want to get the reputation for being a difficult person. If you and your mentor(s) decide that you have a flawed review, you should contact the program officer and determine whether they have any insight into the problem and request their guidance in proceeding with an appeal. You should write a letter, presenting point by point each item in the review that is incorrect or biased. Regardless of how upset you are, your letter will be more successful if you craft a careful and unemotional response. You must recognize that the appeal process frequently results in an additional cycle, as opposed to simply responding and resubmitting.

Seriously Flawed Reviews Can Be Appealed Successfully

Sam's mentor suggested that she apply for a postdoctoral fellowship grant. When the review came back, it was full of inaccuracies and very vindictive. The inaccuracies included statements that Sam's mentor had not produced academicians, although Sam's mentor had three former postdoctoral fel-

> *lows and five former graduate students on faculty in leading institutions.*
> *A reviewer also claimed that Sam's mentor was no longer a productive re-*
> *searcher, even though Sam's mentor had NIH support and had produced 10*
> *peer-reviewed publications last year. The reviewer did not mention Sam's*
> *Ph.D.nor her M.D. Sam and her mentor wrote a letter requesting another*
> *review, in which they documented the inaccuracies. Upon review, Sam's*
> *postdoctoral fellowship was funded.*

In nearly every case, your review will not be flawed. Regardless of how many grants you apply for, it still is devastating to be turned down. Too many applicants give up at this point. They turn over control of their research program to a review group. They forget that, just as it is necessary to invest time to master a method, it takes time to learn how to write grant proposals. The writing skills for grant proposals are honed by responding to the critiques of reviewers.

Loss of Self Confidence Is One of the Most Serious Impediments to Success

> *Sam and Sandy each applied for an NIH R01. Both of their applications*
> *were triaged, but each of them received constructive criticism. Sam was in-*
> *capacitated by the rejection and was so angry that he could not make use*
> *of the feedback. Sam's mentor urged him to amend the original application*
> *and resubmit, but Sam could not bring himself to rewrite the proposal.*
> *Sandy was not pleased when her grant was not funded, but she decided that*
> *the only way she would have support for her research was to complete an*
> *amended grant proposal. Sandy did not succeed with the first resubmission*
> *but the second amended application was funded. Sam decided that he was*
> *not cut out for research and emphasized other aspects of his career.*

Resubmit at the Next Funding Cycle, if at All Possible

Your strategy for amending your application should be to submit your amended application for the next review cycle. You should have additional data and publications for this revision. Your responses to the opinions of the reviewers should be rational and, preferably, data based. You should simply accept the proposed budget cuts and seriously consider any decreased duration recommended by the Study Section. Even if you don't accept their suggestions, they will impose them on the amended application. There is no point in antagonizing the members of the review group. You should correct the factual mistakes of the review in a calm manner. The important suggestions in the review need to be dealt with in the introduction section of your revised application. Any change from the original proposal needs to be in a different type font, italicized, or underlined. Bolded changes may be permissible but they do not duplicate well. In addition to dealing with the criticisms, you must show progress since the original application.

Show Progress toward Your Specific Aims in the Interval between Your Original and Amended Submissions

Kim's initial proposal was not funded, but the feedback was very helpful. By carefully reviewing the critique, Kim determined that most of the criticisms could be dealt with by completing several experiments that would provide the answers to questions raised by the reviewers. With hard work and a little luck, Kim was able to complete the experiments before he submitted the amended application. The reviewers were very impressed with Kim's progress since the original submission and his responsiveness to the Study Section's original review. Kim's first amended application was funded.

Experienced Investigators Frequently Must Resubmit Applications, Too

For any proposal, you may submit two amended applications within three years of your original submission. If you were funded previously and, after an initial unsuccessful review, you requested an unfunded extension to continue the same grant number, you recompete as a renewal and not a new grant. If you revise your proposal in accordance with the reviewers' suggestions and have made significant progress on your specific aims, you should improve your percentile on the amended proposal. Hopefully, this will put you in the fundable range. If you are still outside the fundable range, you should submit a second amended proposal revised in accord with the suggestions made by the reviewers. Certainly, we would all like to have our initial proposal funded. However, the competition for research funding is so intense that even experienced investigators sometimes have to submit one or two amended applications in order to receive funding for their research.

If You Are Unsuccessful in Three Attempts, Develop a New Strategy

Chris' initial submission was triaged. Chris carefully prepared an amended application and received a 35th percentile rating. This was definitely an improvement and Chris was encouraged. Chris' second amended application was at the 25th percentile. Unfortunately, neither institute that Chris requested funded to that level. Chris could not submit this proposal a fourth time and decided to redirect her research efforts based on a collaboration with a colleague. While this research effort was still in the same field, it was sufficiently different to be recognized as a new proposal.

chapter

7

· · · · · · · · · ·

PREPARATION OF ABSTRACTS FOR SCIENTIFIC MEETINGS

· · · · · · · · · · · · · · · · ·

PRESENTATIONS AT SCIENTIFIC MEETINGS ARE CRITICAL FOR THE FOUNDATION OF YOUR NETWORK AND YOUR CAREER

You need grant support to guarantee funding for your research program or even for your salary. Only at the beginning of your academic career as a trainee do you receive grant funds based on the ideas in your proposal and your potential. As your career advances, the basis of your research proposals will be your preliminary data as evidenced by your raw data or, preferably, your publications, in addition to your ideas. Peer-reviewed abstracts and publications indicate that others have considered your ideas and your data and found them to be substantive. Previous success in review of your work for meeting presentations and, more importantly, publications, lends credibility to your proposal and your commitment to see your research program through presentation to colleagues at meetings to completion in the form of publications. Your proposal is an attempt to market your research ideas and so is submitting an abstract to a scientific meeting. Your presentations at scientific meetings provide the basis for your publications and grant proposals and help you to establish your scientific network.

Know the Goal for Your Presentation as You Prepare Your Abstract

Your goal in preparing your abstract is to present your ideas to your colleagues. In order to do this, your abstract must be selected for presentation. Once your abstract has been selected for presentation, you also want it to attract an audience. Each of these goals is part of developing respect for your research among your colleagues.

Good Science Requires Good Writing to Be Effective

The most important requirement for a successful abstract is good science. However, good writing, appropriate for the meeting, is also important. Sometimes, good writing may give the impression that there is more substance to the science than there is, and you should be cautious not to promise more than you have "in hand" and, therefore, more than you will be able to deliver at the meeting. Abstracts should be clear and logical. You will need to include sufficient data to support your conclusion.

A Successful Abstract Is Focused and Understandable

There should be one idea or, occasionally, two very closely related ideas that you can sell as one. More than one idea can be used to develop multiple abstracts. You should never use one idea as the basis for more than one abstract. In addition to being focused, a successful abstract is clearly written without excessive jargon or abbreviations. It should be understandable in a single reading by someone who is unfamiliar with your work. You should also conform to any size and font restrictions to produce a neat, readable abstract.

A Good Abstract Must Have One Purpose, No More and No Less

Lee is a postdoctoral fellow and Lynn is a medical student. They worked together on the same, successful project in Dr. Smith's laboratory over the summer. Dr. Smith wanted the work presented at a national meeting, and wanted both Lee and Lynn to have the opportunity to present their work.

Dr. Smith split the single project into two abstracts, one with Lee as first author and potential presenter and the other with Lynn as first author and potential presenter. The abstract reviewers were not impressed with these "minimal" abstracts. Lee's abstract was not accepted for presentation and Lynn's abstract was accepted as a poster. Dr. Smith then realized that one abstract should have been submitted with the request that it be presented as a poster. Then both Lee and Lynn could have presented their work. When Dr. Smith realized this error in judgment, Dr. Smith asked them to present the poster together.

COMPONENTS OF AN ABSTRACT

Some organizations require structured abstracts with headings for each section. Even if this is not the case, you should provide a tight and explicit structure for your abstract.

Title

Your title should be concise and unambiguous. It should represent a "miniabstract" of your work since it will create the first impression of the abstract for reviewers and, if accepted, for attendees. You do not want the title and authors to occupy an excessive amount of the total space allotted for your abstract. Your title should be forceful and should begin with an important word. Never (or hardly ever) begin a title with "A" or "The." Your title should include your independent variable, your dependent variable, and the population or species studied. It should be provocative enough to attract an audience. Mark Schuster, M.D., Ph.D., had an abstract title that garnered a lot of attention: "Sexual Practices of Adolescent Virgins." This study showed that adolescents were engaging in behaviors that put them at risk for acquiring sexually transmitted diseases in the absence of what many of them considered to be sexual intercourse. The apparent internal contradiction encouraged the reader to go from the title to the substance of the abstract. You may wish to make a declarative statement in the title, but do not overstate your conclusions.

Authors

The authors should include only those who made substantive contributions to the research. The first author does most of the work. The last author is usually the senior person providing overall direction. Helpful colleagues who are not major contributors to the project can be mentioned in the acknowledgments section of your manuscript. Authorship of the abstract is usually the same as that for the manuscript, unless the manuscript combines more than one abstract, new data have been added since the abstract, or the first author on the abstract does not prepare the manuscript. By agreeing to the order and listing of the authors on the abstract, each author is endorsing the authorship as specified.

Agree on the Order of Authorship at the Outset of the Research

Dr. Jones was included as a collaborator on an abstract prepared by Sandy in Dr. Smith's group. Sandy was the first author and Dr. Smith was the last author. The work was accomplished by Sandy in Dr. Smith's laboratory. Dr. Jones provided an essential reagent for Sandy's work. All the collaborators approved the abstract. When Sandy prepared the manuscript based on the abstract, she used the names and order of authors from the abstract. Dr. Jones objected and claimed that he deserved equal status with Dr. Smith. Dr. Jones wanted either first or senior authorship and threatened not to sign the transfer of copyright form if he did not get his way. Dr. Smith decided it was more important for Sandy to be first author, allowed Dr. Jones to be last author, and he became the subsenior (next to the last) author. Dr. Jones agreed to this. Sandy objected and stated that Dr. Smith should be the last author. Dr. Smith explained to Sandy that Sandy deserved the credit for doing the research and preparing the manuscript. Dr. Smith also stated that they would not collaborate with Dr. Jones again.

Introduction

Your introduction should provide the context for your investigations and indicate why you are interested in this area of research. You should provide insight into the foundation for the research problem in one, two, or (rarely) three sentences. This should be written to show that your research was the next obvious step, given the previous state of knowledge.

Purpose

The purpose indicates why this research was undertaken. It provides a statement of your hypothesis in one sentence and may be considered the specific aim of your work. Make the structure of the abstract clear by stating something like, "The purpose of these investigations was. . . ."

Methods

The methods section includes your experimental design. You should specify the animals or population studies, your control group(s), and how they were selected. For standard techniques or treatments, you need not specify the details. If you developed a new technique, then you need to provide more specifics. You should include any statistical approaches you may have used for data comparison and analysis. Two to three sentences should be allocated for the methods section. You may wish to initiate the first sentence with a statement like, "The Methods we employed included. . . ."

Results

In the results section, you will summarize your data thoroughly but very succinctly. Whenever your data allow, your presentation should be quantitative, not qualitative. Be cautious about using tables and figures in your abstract. If you use a table, you still need to describe the data in prose, and you probably will not have room to do both. There is never enough room in an abstract for a figure. That is, we have never seen one presented successfully. It is always too small to be readable. Results can be summarized in two to three sentences. Again, the section can be introduced with a statement alerting the reviewers and readers regarding its content, such as, "The Results of our studies included . . . ."

Conclusions

Your conclusion ("We conclude . . .") should state your main point responsive to your purpose in one sentence. Do not include in your conclusion a promise of more data at the meeting or reach conclusions supported only by preliminary results. Your initial results may not hold up to repetition, or you may not be able to accomplish the research necessary to permit acquisition of the additional data you promised. In either case, you may have to withdraw your abstract, which is very embarrassing and will lead others to question your scientific credibility. If you have data in which you are confident, present it, because abstracts are selected on the basis of data, not promises.

Speculation/Recommendation

In one sentence of speculation ("We speculate that . . ."), you can present the implications, importance, and future directions of your research. If there are practical implications for your results, then present these as a recommendation ("We recommend that . . .") describing how your work may alter current practice.

SUGGESTIONS TO IMPROVE THE CLARITY OF YOUR ABSTRACT

Give Yourself Adequate Time for Preparing Your Abstract

You should never wait until the day an abstract is due to begin writing it. You need to allow adequate time for multiple drafts and for your coauthors to review your near-final draft. Fax or e-mail the abstract with the message that if you do not hear from your coauthors by a specified deadline (at least 24–48 hours), you will assume that they have no changes.

Organize Your Time Effectively for Successful Abstract Preparation

Lee always says he works well under pressure. He likes to wait until the day the abstract is due to write it so that he will have as much data as possible to include. Not only does Lee wait to write the abstract, but he also postpones reading the instructions until the day it is due. Lee is submitting three abstracts to a national meeting. The first abstract covers research performed with Lee's graduate student. Lee would like to nominate his graduate student for the student research award but does not have time to write the letter of nomination. The second abstract involves a coauthor in Europe. By the time Lee has the abstract written, Lee's European colleague has left work for the day, and Lee does not have her home telephone or fax number or e-mail address. Lee decides to submit the abstract without showing it to his European colleague. Later, Lee has to withdraw it when his European colleague objects to submitting an abstract with very preliminary data that have not been confirmed. The third abstract appears to be relatively straightforward, but Lee does not do a good job of preparing it. It is poorly written, contains several errors that confuse the reviewers of the abstract, and is not accepted for presentation. Lee should have taken more time to review and polish each of these abstracts, and he should have been well enough organized to nominate his graduate student for the award.

You Want Your Abstract to Be Strong and Hard-Hitting

Use key phrases to indicate structure in unstructured abstracts: "The purpose"; "We studied"; "We found"; "We conclude"; and "We speculate." Own your research. Use the first person with the active voice: "We observed" instead of "It was observed that." You will find that it takes fewer words. You should use the present tense for previously published results or for conclusions and speculations that extend beyond the present study. Use the past tense when referring to results of the current study or when citing the authors' previous studies.

Abstracts Must Be Clear and Easy to Interpret

You should avoid abbreviations unless they are standard, and even those should be kept to a minimum. Scientific jargon makes an abstract difficult for reviewers or readers who are not in your particular field to comprehend.

SUBMITTING YOUR ABSTRACT

Read and Follow the Directions

Follow the instructions for submission. Your abstract can be rejected if you fail to do so. If you are typing your abstract onto a form, use the largest typeface possible. Typically, abstracts are reduced 70%. To test the readability of yours, use your copy machine to reduce your abstract and consider the result. If it is not readable, you will need to change the typeface. Remember, you cannot market your work if your abstract is unreadable.

Electronic Submissions Require Advanced Planning and Submission

While electronic submissions ease the burden of typing, they have their own problems. Systems are often overwhelmed close to the deadline since many applicants have procrastinated. Whereas deadlines are often extended under these circumstances, you can spend a lot of time being frustrated while you try again and again to submit your abstract. You also need to be sure that you receive a confirmation of receipt of your electronically submitted document.

Avoid Bad Abstracts

You want to avoid the problems of bad abstracts. You do not want your abstract to be unreadable due to bad writing, a typeface that is too small, abbreviations, jargon, or unreadable figures. You do not want your abstract to lack substance because you have too few subjects (avoid case reports that contribute nothing new), your data are too preliminary, or you just make promises. Promises may not be fulfilled, causing you to withdraw the abstract or be embarrassed at its presentation.

Build Your Career on Quality

You want to be sure that your abstract is new, important, and exciting. Even if you have your heart set on presenting your work at a particular meeting, if it is not ready, you should defer until it is. Take the long view of your career and emphasize quality. It takes years to build a reputation for strong careful science, but it is very easy to destroy such a reputation with an abstract that has to be withdrawn or one that is lacking in substance.

Timeline for Abstract Preparation

Activity	Three Months	Two Months	One Month	Two Weeks	One Week
Select topic	X				
Complete research	X	X	X		
Prepare abstract		X	X		
Send abstract to coauthors				X	
Revise abstract					X
Submit abstract					X

Time before Deadline

8

PRESENTATIONS AT SCIENTIFIC MEETINGS

Preparation of Effective Slides and Posters

OVERVIEW

Prepare a Professional, Polished Presentation

When you consider the amount of effort and funds you expended in producing the results summarized in your abstract that was accepted for presentation to a scientific meeting, you should not scrimp on the cost of materials or on the effort required for an effective presentation. We have all witnessed less than professional presentations by established scientists as well as by beginning trainees. A little planning and preparation can lead to a polished presentation.

Slides Should Be Tailored to Your Presentation

Lee was excited to have her abstract accepted for a platform presentation at a national meeting. She had a number of commitments before the meeting, including preparing and teaching a new course. Lee had a collection of slides she could use as the introductory material for her presentation.

She prepared a few new data slides and brought the relevant slides along to the meeting. Lee selected and arranged the slides on her flight to the meeting. While she was doing this, she realized that it would have been better to prepare an entirely new set of slides. A number of the slides contained too much information, and others had spelling errors or information that was not pertinent to the presentation. Some of the slides were in color, others were black letters on a white background, and overall, they did not flow smoothly through the talk. Lee's presentation went reasonably well, but she vowed to prepare a new set of slides for her next major talk.

EFFECTIVE SLIDE PREPARATION

Follow the "Rule of Sevens" to Produce Effective Slides

The most common errors in slide preparation are too much writing, too much information, and too small a type size. You should be able to hold your slides at arm's length and read the words. Consider the "Rule of Sevens" when making your slides: a slide should be no more than seven lines long and seven words wide. Use the full rectangular space of the 35 mm slide format.

Large Rooms Require Even Greater Care in Slide Preparation

If you will be talking to several thousand people, you may need to make new slides with an unusually large typeface. You do not want your audience distracted while they are straining to read type that is too small. Not only will they not be able to read what is on your slide, they also will not be listening to what you are saying. For large group presentations, find out whether you are expected to have duplicate slides for dual projectors.

Your Slides Should Appear to Go Together

You should consider preparing all new slides with the same format for an important talk. The title for each slide should be in a larger font than the text or all in capital letters, with a color scheme or format that emphasizes the title, such as a light color on a dark background, and set off in some manner from the border of the text. We use yellow letters on a very dark blue or black background. The text should be in upper and lower case in a different light color on a different dark background. We use white letters on a dark blue background. Do not use red letters or red line drawings because the audience frequently will not be able to see them. Red may be used if it is shaded toward orange or pink, but still use caution. If you make your own slides using a software program, learn the differences in color and size between what you see on the screen and what appears on the slide. Use shading to give depth to your letters. A horizontal format is preferred because vertical slides often extend above or below the screen. Your text should be bullets with phrases, not paragraphs. Your talk can add information to that provided on your slides.

Your Abstract Forms an Outline for Your Presentation

Review your abstract and use it to formulate your slides. If it was written according to the format described above, it will contain the sections relevant for a 10-minute talk. Use your abstract as the outline and expand with additional content or new information. A 10-minute talk will usually encompass 10–15 slides. The old rule was one slide per minute, but we have found that, as we put less information on each slide, we use 12–15 slides for a 10-minute talk.

Avoid Too Much Information in Your Tables and Figures

We have all seen slides with so much information that they were incomprehensible. Tables should contain no more than two columns by four rows or three columns by three rows. Tables from journal articles have too much information and should not be used. Instead, you can often summarize the important information in a new table. Bar graphs should have no more than six bars. It is important to label the axes. Pie graphs should include numeric percentages for each part of the pie. Most line graphs should have a zero origin for both axes, the axes should be labeled, and rarely should the equation for the line be a part of the graph.

Know the Rules and Conventions of the Organization to Which You Are Presenting

For example, one national organization banned slide(s) acknowledging a presenter's collaborators, arguing that they were taking time away from the science and were often redundant, with the coauthors being named on the abstract. Collaborators not thus named could be acknowledged in the publication of the work. There are also cultures specific to different organizations. Use your network of colleagues to learn the culture.

EFFECTIVE POSTER PRESENTATION

Prepare a Poster with a Professional Appearance

Your poster can be prepared in the traditional format, gluing white paper onto colored poster board, or you may want to laminate your poster. If you live in an area with high humidity or you are presenting in an area with high humidity, you may find that glue does not stick. With computers, you can print your poster on a single large piece of paper and simply roll it up and transport it to the meeting in a tube. This, too, can be laminated. Be sure to conform to the dimensions specified by the meeting organizers. Most poster formats are horizontal rectangles. Bringing your own pushpins can also be helpful.

Format Your Poster to Make Its Organization Logical and Sequential

The heading runs along the top of your poster. The letters should be at least one inch tall. The heading includes the title, authors, and authors' affiliations. Your institution's logo adds to the attractiveness of your poster.

A copy of your abstract should be placed in the upper left corner below the heading. Place subsequent sections in a column below the abstract. Additional material should go into another column. Should you have a number of people at your poster at the same time, the columnar organization allows several people to read your poster at once without getting in each other's way. Each section of your abstract should be expanded and labeled to provide introduction, purpose, methods, results, conclusion, and speculation. Emphasize your results with figures, graphs, and tables. Use color to attract attention, but do not overdo it. Each figure, graph, or table should have its own legend. The print for the body of the poster should be readable at a distance of at least three feet.

Remember That Your Poster Is Not Totally Comprehensive

Your poster should be used as an outline for more in-depth discussion with those who come to view it. In addition to a well-written abstract, visitors will be attracted by your creative use of color and your overall visual presentation. Using computer programs, you can produce posters that have the same colors as your slides, for example, white letters on a dark blue background.

Your Poster Provides an Opportunity to Interact with Colleagues

You should display your poster and remove it in compliance with the instructions of the meeting organizers. You should also be present during the entire viewing time for your poster. If you are presenting more than one poster at the same time, one of your colleagues should be with the other poster(s) at all times. Be sure to bring your business cards to distribute to contacts you make. These are especially important if you wish to receive reprints, reagents, or clinical material from contacts you make at the meeting. You may also consider bringing copies of your most recent, relevant article to give to interested parties.

In the past, scientists considered posters to be less desirable than platform presentations. However, posters provide an opportunity to meet scientists in your field and to discuss your work with them. Often, very senior scientists that you may not feel comfortable approaching will come to your display. Mentors should attend the posters of their trainees or junior faculty members in order to introduce them to colleagues. However, the junior person should answer questions and engage in any discussion, with the mentor only providing support to the junior person.

One colleague talks about meetings in general, and posters in particular, as providing "face time." Present your best work effectively and use your talks and posters as opportunities to network.

9

· · · · · · · · · ·

THE 10-MINUTE TALK

· · · · · · · · · · · · · · · · ·

MAKE SURE YOUR AUDIENCE REMEMBERS ONE THING

Your goal in a 10-minute talk is to leave your audience remembering one important "take-home lesson" from your presentation. Only your closest colleagues and fiercest competitors will remember more detail, so focus on making your key point effectively. Remember the adage: "Tell them what you are going to tell them, tell them, and tell them what you told them."

YOUR SLIDES ORGANIZE YOUR TALK

In preparing your talk, you should make your slides first and then practice or write your talk to explain your slides. There are two very different points of view regarding reading versus talking in your presentation. Decide which format is most comfortable for you. Regardless of the format you choose, try to have your slides ready two weeks before your meeting so that you can practice and revise them as suggested by your own review and that of your colleagues.

A 10-MINUTE TALK SHOULD ACTUALLY BE 10 MINUTES LONG

A 10-minute talk is very difficult because you have so little time. The organization must be very tight without overwhelming the audience with too rapid a delivery or excessive detail. Practice will establish your rate of presentation and help you to determine whether or not you have the right number of slides and the right slides. Remember, 15 minutes have been allocated for your presentation. If it takes you only 5 minutes to present your talk, you will have 10 minutes for questions.

Practice for the Questions after Your Talk

The most important aspect of practice is the questions asked by your colleagues. Too many presentations are derailed when the presenter is unable to answer the simplest question. You need to have colleagues and mentors who will ask you the difficult questions at home and help you to formulate your answers. If you anticipate an especially aggressive rival in the audience, one of your local colleagues should assume the rival's persona and challenge you with the questions you expect from such a rival. Practice answering questions succinctly in "sound bytes" of no more than three sentences. If the answer is not thorough enough, the questioner will follow up with another question. Answers that are too long will cause the audience to lose interest. If you get unexpected questions at the actual presentation, then your practice sessions were not sufficiently thorough.

Bad Answers Will Be Remembered Even if the Talk Is Good

Lynn is a graduate student preparing to give his first 10-minute talk at a national professional meeting. Lynn's mentor assists him in preparing his slides and organizing them for the talk. The department chair always organizes a practice session for those selected to give presentations at the meeting. Lynn practices his talk at this session, and it goes very well. The audience discusses Lynn's presentation in general, and then Lynn goes back through his talk one slide at a time, with the audience making appropriate suggestions for changes. With the feedback Lynn receives, his presentation is greatly improved. Lynn makes a flawless presentation at the meeting. However, when the questions begin, Lynn realizes he did not truly understand his project. The audience soon realizes this as well when Lynn stumbles over his answers. Some of the answers take more than one minute, during which Lynn talks around the topic but never answers the question. Other answers are only a single word, "yes" or "no."

Answering Questions Is a Learned Art

Lee's mentor understands the importance of practicing not only the talk but also the questions. Lee practices in front of her lab group. In addition to providing feedback on Lee's talk, her group asks more than 10 questions. Some of these are very difficult, so Lee has the opportunity of answering them and receiving feedback. Lee also receives suggestions regarding her style of answering questions. Lee is taught to briefly restate the question before answering, since most of the audience will be unable to hear the question. This also gives her more time to consider her answer, which is especially helpful if the question is difficult. Part of Lee's training includes the admonition that it is appropriate to say that she does not understand the question and to ask for clarification from the questioner or the chair of the session. In addition, Lee can admit to not knowing the answer if that is the case. She knows to keep her answers brief, no more than two or three sentences. If asked about work in progress, Lee will be ready to respond that "This is an interesting question, and we are currently pursuing this line of research."

THE NUMBER OF SLIDES FOR YOUR PRESENTATION SHOULD BE PLANNED AHEAD OF TIME

A general rule is 1½ slides per minute, or about 15 slides for a 10-minute talk. If the slide is a simple word slide or photograph, you can expect to spend 30 seconds on it. If the slide illustrates a complex procedure or theory, plan to spend 2 minutes.

Your talk is essentially an expansion of your abstract. You should have one slide with title, authors, and affiliations, if it is permitted. Some organizations do not allow you to use such a slide since the session moderator reads the title, names of the authors, and affiliations when the speaker is introduced. There should be one or two slides for the introduction. The purpose should take one slide. Between one and three slides can be used for the methods. Three to five slides are needed to present the results. If you use summary slides, there should be only one or two. Conclusions can be covered in one or two slides. If you have any speculation or recommendation, use one slide each.

BE SURE TO ACKNOWLEDGE THE WORK OF OTHERS

In planning your slides, consider the background of your audience. If your slide comes from the written work of someone else, you should provide a citation at the bottom of the slide and acknowledge the source in your talk. You do not want to give the impression that you are taking credit for work done by someone else.

PREVIEW YOUR SLIDES AND THE PODIUM, AND BE GRACIOUS DURING AND AFTER YOUR PRESENTATION

When you load your slide tray in the speaker ready room, look through them to be sure that they are in the correct orientation. Go to the room where you will make your presentation before the session begins. Check out the podium and equipment. Make sure you understand how to advance the slides and how the pointer works. Meet the moderator and the projectionist. Speak slowly during your presentation. We all talk faster when we are nervous. Thank the moderator and projectionist before you leave. If there are problems with your slides during your presentation, maintain your equanimity. Your audience will remember you if you handle an unfortunate situation with grace, and they will also remember if you do not.

If you have a questioner who is arrogant, aggressive, or naive, handle yourself with poise. The audience will be on your side and will respect you if you respond evenly and effectively. If the questioner is particularly obnoxious, you have no need to respond—you have not entered into a social contract with them that demands a response. Just thank them and move on to the next questioner.

10

··········

THE 1-HOUR TALK, INCLUDING THE JOB APPLICATION SEMINAR

··················

THE 1-HOUR TALK

THE "1-HOUR TALK" IS REALLY 30–45 MINUTES LONG

First, consider the length of the "1 hour talk." It is not 1 hour long. It will start 5 to 10 minutes late, and you should leave at least 10 minutes for questions at the end. So, it is closer to 40–45 minutes in length. If you do not know the culture of the institution, you may wish to ask your host. Some audiences will begin to leave 10 minutes before the hour because of clinical or teaching obligations that begin on the hour. For this latter group, your presentation should be only 30–35 minutes long. And we have had the experience of anticipating a 1-hour talk in a foreign venue and learning upon arrival that a 3-hour talk was anticipated. In foreign settings, clarify ahead of time the type of slides you will be using and the capacity of their slide trays. There may be only 30 slides per tray.

Know the Audience for Your Presentation

As you prepare your 1-hour talk, consider the venue. You should ask the person inviting you about the audience and the type of talk they are expecting. Is it scientific or clinical? Will it be an educational or a peer presentation? Are the members of the audience from the same specialty or subspecialty in which you practice? Will they be a homogeneous group or a group with mixed backgrounds and knowledge bases? Is this a local, regional, national, or international meeting?

You need to consider the sophistication of your audience and be sure that you define any terms that may be unfamiliar to them. Another variable is the interest level of your audience. Are they attending the meeting for intellectual curiosity, educational purposes, or recreational purposes? If it is a meeting in a destination location, you may have to be particularly creative with your title to get the audience into your talk and bold in your presentation to keep them. There are different types of 1-hour talks, including lectures, seminars, and problem-solving sessions. You need to determine which type you are to present, because the goals and formats will be quite different.

Tailor Your Talk to Your Audience

Kim always gives the same talk in every setting: a very sophisticated presentation of his research, appropriate to scientists in his field. In doing this, Kim misses the opportunity to educate nonscientists, trainees, and scientists in other fields. It also costs him audience members. When Kim comes to a university, those who heard the same talk at a recent national meeting find better ways to spend their time than to listen to the talk for a second time. Eventually, invitations to present talks at professional meetings or seminars at other universities decrease. Kim's lack of enthusiasm for his audience as projected by his lack of interest in developing an appropriate presentation has lead to a lack of enthusiasm for inviting Kim to speak.

Your Title Should Be Interesting, Your Organization Clear, and Any Potential Conflicts Apparent

As you prepare your slides, you need to make a title slide. Your title should be interesting to your potential audience. You should never use the title of your last paper. You may want to include subtitles to help orient your audience. You need to acknowledge the source of support for your research. This is especially important if the money comes from industry and there is a potential conflict of interest, and it is also important if representatives of the funding agency supporting your research will be in the audience. Determine whether you will be asked to provide an outline, learning objectives and/or a manuscript.

Be Cautious of Giving Up Copyright to Your Presentation

If these materials will be published in any form, including on the Internet, you will no longer own the copyright to the material. You do not want to provide for this publication any new data that you plan to publish in a peer-reviewed journal, since you will be unable to publish this new material in an original article if it has already been published. If copies of your slides are to be published on paper or on the Internet, the copyright on your slides will be owned by the publisher and using your slides in future presentations may be a violation of this copyright.

Maintain Control of Your Slide Preparation

Occasionally organizations will offer to make your slides for you. This is tempting as a money-saving strategy. However, there are problems. Unless you inquire specifically, they may not tell you that they plan to publish your slides. They may distribute printed copies to those attending the meeting. They may also use copies of your slides on tape, on a CD, on the Internet, or as slides. You would be wise to decline the offer to have the organization make your slides. As part of the service of making slides, the organization may also offer to have your slides ready at the meeting for you, but you should bring your own copies. There have been several meetings where the organization did not bring the slides for one or more speakers, or they were lost in shipment. Without their own set, the speakers had no slides for their talks.

Inquire Carefully about Copyright Ownership for Your Slides and Presentation

Sam was invited to give a talk as part of a continuing education course for professionals. The course was run by a prestigious national organization. Sam was flattered to be invited. She was planning to use slides she had developed over the past five years for a lecture she presented several times a year to trainees. Sam was concerned when the organization wanted her slides on disk so that they could prepare the slides for her talk. When Sam called to speak to the course coordinator, the coordinator said the organization wanted the slides to be all the same format. Sam had never been to a conference or course where this had been the case. When Sam asked the coordinator what was the motivation for the uniformity of slides, the coordinator admitted that they planned to print the slides of all speakers as a book for the course participants, for sale to others and for use on the organization's Web site. Sam asked who would own the copyright if she sent in a disk containing her slides. The coordinator stated that Sam would own the copyright. Fortunately, Sam did not accept this response. She suggested that the coordinator check with the organization's lawyers. Three days later, the coordinator contacted Sam, stating that the organization would own the slides. Furthermore, if Sam subsequently used the slides in her teaching, it would be a violation of the organization's copyright. At this point, Sam almost said that under the circumstances, she would not be able to participate in the course. However, she did believe in the purpose of the course and felt that it was important to the profession. Sam told the coordinator she would bring her own slides to the meeting and provide an outline for the coursebook. The organization put off preparation of the slides for the other speakers until the last minute. They were lost by the overnight courier and did not arrive at the course on time. Sam was one of the few participants to bring their own slides and also one of the few able to present their talks with slides. The meeting was a disappointment for the speakers and the participants, and the organization was embarrassed by the course. Sam was the only speaker who knew that the organization held the copyright to the slides produced, copied, and marketed by the organization. The organization never informed the other speakers.

Do Not Trust Others with Physical Control of Your Slides

At some meetings, the organizers want to take your slides once you have loaded them in the slide tray and deliver them to the meeting room for you. There have been numerous times when the slides did not come to the correct room. Regardless of the suggestions of the organization running the meeting, make your own slides and carry your slide tray to the presentation yourself. If the organization objects, you can decline to make the presentation. If the slides do not show up, it reflects on your professionalism. You have worked too hard to let an organization own your work or interfere with the presentation of your work.

Organize Your Presentation Like a Banquet

As you plan your talk, think about the last banquet you attended. Think of the introduction and summary as the appetizer and dessert. In between, you can accommodate three to five courses. If you eat more than that, you would be unable to remember what you ate. If you have more than three to five points, your audience will be unable to remember them.

Plan Your Presentation Like a Script or Screenplay

Remember what maintains an audience's interest in a play or movie. Plan your talk to include changes of pace. If you can organize your talk to create suspense, you should do this. For example, if you have been successful in answering a long-standing question—solving a mystery as it were—then build to the climax, perhaps even including the occasional wrong turn that you made. The audience will bond to someone who, like themselves, is imperfect.

Preparation Will Reduce Anxiety

In order to deal with your anxiety, you should practice at home in front of critical colleagues who will ask you difficult questions. Be sure that your slides are correct, that your practice audience can follow your organization, and that your presentation fits within the assigned time. Preview your slides in the speaker ready room at the meeting and take them to the presentation room before your session begins, so that you can check out the podium and equipment and introduce yourself to the moderator and projectionist.

Remember to Breathe

At the time of your presentation, you may want to have a glass of water handy. Taking a sip of water will help you to catch your breath if you are nervous. You need to remember to breathe when you are speaking. Select several individuals in the audience and make eye contact with them. It will help the audience bond with you, and it makes a large audience seem smaller. The first three to five minutes are the most difficult. You should visualize your presentations that have gone well, and recognize that most of your fears are not based on any prior experience you have had as a speaker.

Enjoy Yourself and Your Audience Will Enjoy Your Talk

Your practice at home should give you confidence and help you relax while you talk. Enjoy yourself, maintaining a sense of humor without trying to be a comedian. Be friendly and enthusiastic.

Remember to Say Thank You so Your Audience Knows When to Clap

At the end of your presentation, you should say thank you. You have been at presentations at the end of which there are embarrassed pauses when the audience is not sure whether it is over. A pleasant and sincere "thank you" is a polite way of letting your audience know the presentation is over and they can applaud.

It's Not Over When It's Over

In your answers to questions, you should be brief, your responses no more than one to three sentences in length. Be sure to repeat the question if the audience cannot hear the questioner. This will also allow you time to plan your answer. Whenever possible, you should attend the entire session to hear other presentations. After all, you wanted an attentive audience for your talk. At the end of the session, you should thank the moderator and sponsor.

Experience Will Give You Confidence

Chris was terrified of speaking in public. As a graduate student, he had had to make numerous presentations to his committee in addition to presentations at scientific meetings. Chris' mentor, Dr. Smith, was very sensitive to his fears. Dr. Smith realized how important presentations were to Chris' success as a graduate student and throughout his entire academic career. Dr. Smith always arranged weekly lab meetings so that each member of the group presented their progress, problems, and plans at every meeting. Chris, like everyone else, gave a weekly presentation of his work. In addi-

tion, Dr. Smith would ask Chris to explain a method, review a paper, or explain an alternative research design weekly. To further desensitize Chris, Dr. Smith had him submit abstracts for regional scientific meetings where there was a good chance of Chris' giving a platform presentation. Dr. Smith made sure Chris was well practiced before he made presentations at scientific meetings. Chris soon gained confidence in his abilities and became an excellent speaker.

THE JOB INTERVIEW

Be Yourself

When you go for an interview, you should be yourself. Perhaps you could fool a potential employer for a few days but not after you accept the position. The worst thing you can do is to get the job for the wrong reason and not satisfy their needs, leaving both you and your new employers unhappy.

Prepare for Your Visit

Before you visit, you should carry out a literature search on those who will be on your agenda. This will enable you to ask appropriate questions. Ask your colleagues and mentors about the people and the place. Make sure you know the kind of talk you are to give and who your audience will be. If someone at that institution has done work related to yours or has collaborated with you, be sure to mention this in your presentation. If the audience is small, don't voice your disappointment. Give a great talk and be as enthusiastic as if you were talking to thousands. Your hosts and the search committee will be there, and they will tell the others what a fantastic talk they missed. Be sure to leave time for questions. Your hosts will want to see how you handle questions, and you should be interested in the questions they have for you.

Be Honest with Yourself and Your Interviewers

Sam could not decide whether she wanted a job in industry or in academia. She applied for academic jobs as if that were her only interest and she approached industry jobs in the same way. On a job interview at a university, the department chair asked Sam how committed she was to an academic career. Sam stated that she was totally committed to academics. The department chair asked Sam why she had interviewed with an industry representative at a recent national meeting. Sam was shocked and did not know how to answer. The chair had a friend working at one of the companies with which Sam had interviewed. When they were discussing their recruits, each mentioned Sam and then realized they were talking about the same person.

Be Nice and Know What You Want

You should be well rested before you go, because the visit will be rigorous with long days scheduled. Even if your hosts are not nice, you should be. Prepare questions ahead of time for the people you are meeting, because people like to talk about themselves and their own work. But do not look at your questions unless you are negotiating with the chair or division head. Know what you want from this position in terms of salary and professional support. They may ask you on the first visit, but take your cues from them. If they don't bring it up, wait until the second visit. You should be diplomatic; do not get involved in internal politics.

A Discussion Is a Shared Interaction—Do Not Monopolize It

Remember that in our culture, a discussion generally represents a roughly even split overall in the time occupied by the verbalizations of each participant. Therefore, beware of monopolizing the conversation, especially if you are someone who tends to talk excessively when you are nervous.

Be Diplomatic during Your Visit

Chris was applying for his first job as an assistant professor. While he was meeting with Dr. Jones, a senior professor in the department, Dr. Jones explained some of the problems Dr. Jones had been having with the department chair, Dr. Smith. A number of the issues Dr. Jones raised were important to Chris. Dr. Jones maintained that Dr. Smith kept all of the departmental administrative resources for herself, that Dr. Smith prevented faculty from traveling to scientific meetings, and that Dr. Smith discouraged faculty from pursuing their own independent research interests. When Chris met with Dr. Smith, Chris was very confrontational and accusatory. Dr. Smith realized that Dr. Jones had met with Chris. Dr. Smith was no longer enthusiastic about Chris' recruitment. The last thing Dr. Smith needed was another Dr. Jones on the faculty.

Remember Who Is Paying for Your Visit

When you are on a job interview, you are at your hosts' disposal. In general, you should not arrange to visit friends or family in the area. If you do, your hosts may feel that you have an ulterior motive and are taking advantage of them for free airfare. If you move to the area, you will have plenty of time to visit. On the other hand, some hosts may know you have family or close friends in the area and may see this as an attraction for you and encourage you to visit them. You should exercise moderation in all areas, including speech (be a good listener), food (a big lunch leads to a sluggish afternoon and a big dinner leads to a big bill; take your cues from your hosts), and drink (for obvious reasons).

Do Not Take Advantage of a Prospective Employer

Dr. Smith invited a postdoctoral fellow candidate, Lynn, to travel to meet Dr. Smith's lab group and to give a seminar. The day before Lynn was scheduled to arrive, Dr. Smith called Lynn to clarify the schedule. Dr. Smith was told that Lynn had left the day before. When Lynn arrived, Dr. Smith inquired about Lynn's travel plans. Lynn had arrived in town two days early to be interviewed for another position. Dr. Smith suggested that Lynn split the cost of the airfare between the two potential job sites. Once Dr. Smith learned about Lynn's attempt to take advantage, Lynn was no longer in contention for the position.

Send Requested Materials Promptly

When you return home, promptly send any materials requested, such as your plan or budget. You should send a thank you note. If spouses or significant others were involved in entertaining you, remember and include their names as well.

Your Spouse/Significant Other Is Part of the Interview Process

Your spouse or significant other usually will be asked to accompany you on the second visit. Your significant other is also being interviewed throughout the visit. You should prepare them by discussing personalities, politics, etc. before your joint visit.

If Your Spouse or Significant Other Is Adamantly Opposed to the Move, There Is Little Purpose to Their Visit

Chris was looking at his first real faculty position, an assistant professorship at a major university. Chris' first visit was very successful, and Chris was invited to return for a second visit with his spouse. Chris' spouse, Lee, did not want to move. Lee especially did not want to move to the area near the university. Lee went to Chris' second visit determined not to have a good time. Although their hosts were very gracious, Lee challenged every positive statement made by the hosts about the university and the surrounding area. Lee found fault with every house they looked at, with every restaurant they ate in, and with every person they met. While the department was still favorably impressed with Chris, the faculty discussed whether they should invest any more resources or energy in him, since his wife was so negative about the move. They decided not to pursue Chris' recruitment, and future professional interactions were somewhat cold because they felt that they had been misled by him.

Get It In Writing and Be Prepared for "Buyer's Remorse"

You should get the offer in writing before you accept. Unfortunately, we have been told by many colleagues of verbal offers that were made and later withdrawn, sometimes with the candidates having made significant life decisions, such as giving notice to their current employer, placing their house on the market, and even selling it. Also, the conditions of the offer may be understood differently from those in the verbal discussion. Remember, as you review the offer letter, however, that no job is perfect, and "buyer's remorse" (concern that your commitment is a mistake) is one of the standard, though transient, consequences when any offer is considered or accepted. Look for the situation that will provide the best environment for professional growth and development.

chapter

11

·········

SELECTING A JOURNAL

Instructions for Authors, Recommending Reviewers and
Submitting the Manuscript

·················

OVERVIEW

Know the Type of Journal You Are Targeting

Before you select a specific journal, you need to consider the type of journal
in which you are interested. Do you want a scientific or a clinical journal? Are
you concerned with sharing fundamental information or do you want to impact on practice? Do you want a general or specialty journal? Would you like
to have a broader impact, beyond the limits of your subspecialty, or does your
work have a specific focus within a particular subspecialty? Do you want to
publish in a new journal or an established one? A new journal may have a
quicker turnaround time, less time between submission and publication.
However, some new journals do not publish in a timely fashion and there is
the risk that they may not survive. In addition, a new journal may not be listed in some of the Internet citation searches, such as Medline, for at least the
first two years of publication.

Pick Three Journals

Once you decide on a specific type of journal, you need to consider the level of prestige in which you are interested. You should select three journals at

three different levels of prestige. First, submit your manuscript to a journal that you feel is extremely competitive and which may be out of reach of your work. Your second choice should be one where you are reasonably sure that your manuscript will be accepted. In case this journal does not accept your manuscript, determine a third journal where you know you are guaranteed acceptance. Having this list of three journals before your initial submission will help protect you from delayed resubmission if your manuscript is rejected by your first or second choice. In developing this list, you are aiming high, because if you do not try for the highly competitive, prestigious journals, then your work will not be published there. Also, you are recognizing that manuscript review is capricious and rejection does not reflect personally on you. This consideration at the outset will help you turn around a manuscript more rapidly if it is rejected.

In addition to weighing prestige, you need to consider the length of time from submission to publication. Look at current issues to determine this turnaround time. Remember that five years from now, June or August of the same year will make no difference, while December of one year is clearly different from January of the next year. You can negotiate with the editor of the journal for an expedited review, whether or not the journal has a policy of expedited review.

When you select your three journals at different levels of prestige, consider any major differences in format. Be prepared to quickly reformat your initial submission to conform with the instructions for authors of your second choice. Keep a copy of the complete references to facilitate this resubmission using a software package for managing references. If you are turned down by your first-choice journal, make any changes you can that are recommended in the reviews, change the format, and submit to your second-choice journal within the week.

Know the Current Editorial Policies for the Journals You Select

When you review the instructions for authors, pay special attention to editorial policy. Can you submit your manuscript to a specific editor who knows your work? Are authors able to recommend reviewers? You cannot recommend those with whom you have personal or professional relationships (e.g., family members, your mentors, or department chair), but you should suggest leaders in your field and those who know your work. Even if the journal does not ask you to recommend reviewers, you can always specify that your mortal enemy should not be allowed to review your manuscript. You do not have to give a reason. Manuscripts have been delayed or rejected by editors or reviewers who are competitors or friends of competitors.

If you cannot name reviewers or specify a particular editor, you should consider the members of the editorial board. Members of the editorial board will probably determine reviewers or be reviewers themselves. You should seek the advice of senior faculty in your field regarding your selection of journals. Scan recent issues of the journal to see whether they publish articles in the format in which you wish to submit. Even if the editorial policy states that they accept review articles, if none have been published in the past six months, you should not submit one. If the journal has recently published an article on your topic, that suggests that they may be interested, but, on the other hand, some journals do not want too many articles on the same topic.

Request Exclusion of Your Nemesis as a Reviewer

Lee and Lynn each prepared a manuscript at approximately the same time, and they submitted their manuscripts to the same journal. Dr. Smith, a mutual nemesis, was a member of the editorial board. Lee requested that Dr. Smith not be allowed to review his manuscript. Lynn made no such request. One month later, Lee was asked to make some reasonable changes in his manuscript and submit a revised manuscript, which was accepted for publication. Lynn still had not heard from the editor of the journal. Lynn spoke to the editor, who said she was having a hard time completing the review process. Lynn reminded the editor that his manuscript had already been under review for three months. When the Editor would not make a specific commitment to Lynn regarding his manuscript, Lynn withdrew his manuscript from consideration and immediately submitted his manuscript to another journal.

Do Not Violate a Prepublication Embargo

Some journals have a prepublication embargo. There cannot be any prepublication publicity about your work if you want your manuscript published in the journal, and this may even include presenting an abstract at a scientific meeting.

Flattery May Tempt You to Violate an Embargo

Kelly submitted an abstract to a national meeting. The abstract was about a very exciting finding. The meeting organizers selected Kelly's abstract as one of the ten best at the meeting. Kelly was invited to meet with members of the press to discuss her research. Kelly was flattered by the selection but had to decline. She had already submitted a manuscript on this work to a journal that had a prepublication embargo. Meeting with members of the press before the article was published in the journal would violate the embargo.

Instructions for Authors

Be Sure Your Information about the Journal Is Current

You should get the most recent version of the instructions for authors. Most journals post these on the Internet. Be sure that you follow the typing instructions and the general and reference formats. Some editorial offices will return manuscripts that do not conform to the instructions, causing unnecessary delays. In addition, if you use the wrong general or reference format, the editors may think you first submitted the article elsewhere and are resubmitting it without careful revision.

Send the correct number of copies of the manuscript. The correct number of labeled figures in the size and format requested should be included. Each figure should be labeled on the back with figure number and first author's name. Place each set of photographs in its own labeled envelope and attach to each copy of the manuscript. In addition, include photocopies of the figures, which you can identify more clearly on their front sides.

Transmit Your Manuscript Properly

When you submit the manuscript, be sure you use the correct address. Some journals welcome electronic or fax submission of manuscripts. This speeds the review process. Other journals will accept a disk, and still others want both a paper and a disk version.

Letter of Transmittal

Remember to Include an Effective Letter of Transmittal

Your letter of transmittal is an opportunity to sell your manuscript to the editor. It should be no more than one page in length. Between the editor's address and the salutation, include a statement: "Re: title of manuscript." Restate the title in the first sentence. You should briefly describe the background to your work, your methods, your results, and your conclusions. Place your work in context and elaborate on any important breakthroughs and implications. You should cite the relationship between your manuscript and any work recently published in the journal.

If the journal requests that you suggest reviewers, do so. Request that your nemesis be excluded from reviewing. You should close with a statement thanking the editor for a rapid review so you end on a positive note.

Follow the Instructions Regarding Copyright

Do you need to make a statement about copyright in the letter of transmittal? Do you need to submit a form signed by each author?

Communicate Effectively and Let the Editors Know How to Contact You

If the journal requests a hard copy or disk version, then submit your manuscript using a reliable overnight delivery system. You will have a receipt that enables you to track your submission, and the editorial office staff will take notice of your submission. If the journal requests electronic submission, be sure that you have acknowledgment of receipt. Include your telephone and fax numbers and your e-mail address so that the editorial office can contact you as needed. You would not want publication of your manuscript held up because the publisher could not contact you with a question.

Maintain Communication with Your Coauthors

Prior to submission, your coauthors need to receive the manuscript with a deadline for their comments. Indicate that if you have not heard from them by the deadline, you will assume that they have no suggestions for changes. When you submit the manuscript, send them a final copy with the date of submission. If you need to revise and resubmit the manuscript, you should send them copies of the reviewers' comments and the letter from the editor along with your revised manuscript for their review. Send them a copy of the article when it is accepted and send them at least five original reprints when they come. Just like you, they need to know the status of the manuscript for their curriculum vitae, promotion and tenure, and for citations in their own manuscripts. It is very frustrating to discover that you are a coauthor on a published paper of which you are unaware. To learn about any papers that were submitted without your knowledge, you should do a literature search for yourself at least once a year.

Journals Lose Manuscripts

A very prestigious journal was producing a special issue devoted to Sam's discipline. Sam submitted a manuscript for the special issue. After the reviews were submitted, the editor said that Sam's manuscript was acceptable but not significant enough to warrant publication in the prestigious journal. The editor suggested that Sam allow the editor to submit Sam's manuscript to a journal published by the same professional organization and affiliated with the prestigious journal. Sam felt that his manuscript would be more likely to be accepted for publication and published more quickly if he allowed the parent journal to forward the manuscript to the related journal. Two months later, Sam contacted the editor of the related journal when there had been no correspondence regarding the manuscript. When Sam telephoned the editorial office, Sam found out that they had never received the manuscript. Somehow, it had been lost between the editorial office of the prestigious journal and the editorial office of the related journal. The editor of the related journal apologized for the delay and expedited the review of Sam's manuscript. If Sam had called the editor to ensure that the manuscript had been transferred two months earlier, he could have avoided this problem.

Keep Track of the Timeline for the Process after Submission

Once the manuscript has been submitted, you should receive an acknowledgment of receipt from the editorial office within two weeks. If you don't, contact the editorial office to be sure that they received the manuscript. If you don't hear from the editor within two months of submission, contact the editorial office to determine the status of the manuscript.

BE A WILLING AND CONSCIENTIOUS REVIEWER

Remember, it is difficult to secure reviewers for manuscripts. When you are asked to review a manuscript in your area, you should do so promptly and thoroughly. Reviewing manuscripts helps you to stay up to date in your field.

It gives you insight into your competitors' research. It can increase the citation of your work if the author has not cited your work and should have. It may also help you find papers where you should be a coauthor.

Sometimes, You May Find You Are Reviewing Your Own Work

Chris was asked to review a manuscript in her area of research. She agreed to review the manuscript since Chris did research on the topic and the authors included some authorities in the field. When Chris read the manuscript, Chris found that the authors included some data that Chris had presented as a poster at a recent national meeting. Chris was furious. There was no reference to her poster. Her initial impulse was to call the corresponding author and demand to know why they had used her research inappropriately. Fortunately, Chris spoke to her mentor before calling the corresponding author. Chris' mentor insisted that this was a matter to be decided by the editor of the journal. When Chris called the editor, the editor asked that Chris fax the abstract and poster from the meeting to the editorial office. The editor agreed with Chris that the data from the manuscript were very similar to those in the abstract and poster. The editor contacted the corresponding author to ask the source of the data. The corresponding author maintained that the data came from the laboratory of a colleague who was cited in the acknowledgments. This acknowledged colleague was collaborating with Chris. When the editor contacted the acknowledged colleague, the editor determined that the colleague had provided Chris' data. The editor rejected the manuscript and warned the authors to be sure of the source of every piece of data. The editor invited Chris to submit a manuscript on these data and suggested that she try to submit a manuscript before presenting the data at the meeting.

12

•••••••••

HOW TO WRITE RESEARCH PAPERS

••••••••••••••••••

CRAFTING A GOOD TITLE TAKES TIME

The title of your research paper is your initial opportunity to attract potential readers. This is the first thing readers see when they scan the table of contents of the journal, do literature searches, and read your curriculum vitae, grant applications, progress reports, and renewals. The title should be a concise statement of the paper's content. There should be enough information in the title so the reader can determine their interest in the article, including the essence of the experimental work, species, source of tissues or cells, and any qualifying phrases (for example, *in vitro*). A declarative title will attract attention, but, in making a brief statement about the core message of the manuscript, be very cautious not to overstate your conclusions in the title. We spend a considerable amount of time developing the title because it is so important for current and future interest in our work. One way to develop a title is to craft a series of statements that declare the key message from your work, trying to focus it sequentially as you mix and match terms and phrases from your previous draft title. Another method is to list the key terms that you would consider for the title and then select from the list to develop a series of draft titles, from which you choose the best. The process of developing a title may stretch over more than one day as you "sleep on" different possibilities.

AUTHORSHIP REQUIRES SUBSTANTIVE CONTRIBUTION

Deciding who should be included as an author, and in what order, is often a point of contention in the preparation of manuscripts. Each author should have made a substantive contribution to the manuscript, not just have performed menial tasks as part of their job. Technicians should be included if they have contributed original thought or extraordinary effort. Physicians providing clinical material should be included if they have provided clinical insight in diagnosing the patient and an intellectual contribution to the research. Your boss, the clinic director, or the laboratory director should not be automatic coauthors if their contributions do not extend beyond their administrative roles; i.e., being the boss is not sufficient for authorship without substantive involvement in the research. In order to obtain independent grant support for your research, you need to have publications without your boss or mentor to demonstrate your actual independence. The acknowledgments section can be used to thank such individuals instead of making them coauthors. The acknowledgments should also include grant support relevant to the performance of this research. You will be asked to submit publications as part of your annual report and your competing renewal of your grant. You should also include any affiliation or support from industry, especially if there is the possibility for the perception of a conflict of interest.

KNOW THE RULES FOR AUTHORSHIP IN YOUR GROUP

Kelly is a graduate student working with Dr. Smith's group. Dr. Smith's group includes a postdoctoral fellow and two other graduate students. Kelly's research is going very well, and in six months Kelly has enough data to write a manuscript. Dr. Smith suggests that Kelly write the manuscript and states that the authors should be in the following order: postdoctoral fellow, Kelly, and Dr. Smith. Kelly asked Dr. Smith why the postdoctoral fellow was even being considered for authorship, since the fellow did not contribute in any way to Kelly's project. Dr. Smith replied that the postdoctoral fellow was generally responsible for the research in the group and, for that reason, always received first authorship. Kelly checked with the other graduate students in the group. The other graduate students said that Dr. Smith had insisted that the postdoctoral fellow be the first author on every paper they had written, and Dr. Smith the last, or senior, author. Kelly went to the department chair, who agreed that Kelly should be the first author.

With the intervention of the department chair, Kelly was made first author of the manuscript, and the postdoctoral fellow was not included as an author. However, both Dr. Smith and the postdoctoral fellow were so upset with the department chair's involvement that Kelly had to find another group in which to work.

ORDER OF AUTHORSHIP SHOULD BE RATIONAL AND MEANINGFUL

The order of the authors can be a contentious matter. The first author should be the person performing the research and preparing the initial draft of the manuscript. In situations where the person doing the bulk of the research work and the individual authoring the first draft are not the same, a decision must be made regarding the relative weight of work and intellectual contribution for these different activities. The last author should be the mentor. Ordering of the other authors should be determined by their levels of contribution. Some papers indicate co-first authorship for two authors, but it must be recognized that this information is lost once the paper is cited or indexed. It might be more meaningful for the two individuals to alternate first authorship if their contributions are truly equal.

THE MENTOR SHOULD INTERCEDE IN CONFLICTS OVER AUTHORSHIP

Kim's group received critical patient materials from another researcher, Dr. Jones. These materials provided the basis for Kim's laboratory research, but Dr. Jones did not provide any intellectual contribution to Kim's research project. Kim prepared the initial version of the manuscript and sent copies to all of the authors. Dr. Jones was furious to be given middle authorship on the manuscript, when Kim was the first author and Kim's mentor, who supervised Kim's research and was responsible for all of the intellectual direction of Kim's work, was the senior author. When Kim's mentor reminded Dr. Jones that Dr. Jones' only contribution was the collection of patient samples, Dr. Jones countered that Kim could not have done the research without the samples and threatened to block publication of the manuscript. Kim's mentor reminded Dr. Jones that the authorship had been established and not challenged in a published abstract. Authorship remained as originally proposed, with Dr. Jones as middle author.

Key Words Are Important to Keep Your Work from Being Lost

Key words are often provided at the last minute. They are very important since the index of the journal and indices of literature searches are based on key words. Spend time making a list that may be too long at first, and then select the best from the list.

Your Abstract Is Second Only to Your Title as an Opportunity to Attract Readers

Each journal will have specific requirements regarding the length, content, and format (e.g., structured or unstructured) of abstracts. Abstracts should be written for a general audience and should briefly describe the background and purpose of the study, methods, results, discussion, conclusions, and speculation, following the guidelines we considered in Chapter 7 on the preparation of abstracts for scientific meetings. You should avoid abbreviations and jargon.

The Introduction Sets the Stage in Three Subparts

The introduction provides the background for your research. It specifies the questions and goals of the research and shows how these developed from previous research. You should write your introduction so that the reader considers your research the next necessary step in the field. It will typically consist of three subsections, though these will not usually have subheadings. The first, usually a single paragraph described as the contextual significance of the work, defines any terms that may be critical to the study. The second subsection is usually one or more paragraphs (rarely more than three) in length and provides the historical background to, and previous literature directly relevant to, the primary topic or purpose of this work. For this reason, this second subsection of the introduction is usually more heavily referenced than the first or third, but it is important not to bring in material that more correctly belongs in the discussion. The third subsection describes the purpose of the study and may include a very brief description of the results and, therefore, is a partial abstract of the work, containing the core message of the paper: its hypothesis (purpose) and brief data summary.

Your Methods Section Should Represent a Logical Experimental Approach

The methods or materials and methods section should be written in enough detail so that anyone can repeat your study. It is best to break this section up with specific subheadings organized according to the design or logic of the study. You should include a description of the relevant attributes of the human participants in your research, including age, sex, ethnicity, and other important characteristics. If the work requires one or more case descriptions, these should be provided in a separate section, usually immediately before or after the materials and methods or as the first subheading in this section, according to the journal's instructions. Patients should not be identified by name or with any information that the individual or family member would find inappropriate or would consider sensitive (e.g., misattribution of paternity). Written consent is required for photographs. Approval by the appropriate institutional review board is required for either human or animal subjects and such approval should be noted. Animals should be described by species, strain, age, sex, and other important characteristics. Refer to other papers for published methods and, if you modified published methods, cite the original and describe the changes you made. You should specify the manufacturer, city, state, or country of unusual chemicals, reagents, and equipment.

Be Cautious about Confidentiality

Kelly was preparing a manuscript on a rare disease. As part of the patient information, Kelly planned to include the family pedigree of a family that was very important to his research. When Kelly reviewed the instructions for authors, he found that the editorial policy of the journal was to restructure family pedigrees so that no one would be able to identify the family member that had the disorder or members who were carriers for the disorder. The rationale was that there may be family members who had not participated in the research or did not want to know their genetic status regarding disease or carrier status. Further, participants in the study may not want others to know their status. For a rare disorder, a family pedigree could identify the family and the individuals within it. To conform to the editorial requirement and to maintain the confidentiality of family members, Kelly constructed an idealized pedigree that conveyed the typical inheritance pattern but would not identify any one family.

THE RESULTS SECTION SHOULD BUILD RATIONALLY TOWARD YOUR LOGICAL CONCLUSIONS

The results describe data summarized in tables and figures and may be organized by subheadings ordered according to the study design. Do not present the same data as both a table and a figure. The statistical measures used should be specified (e.g., standard deviation or standard error of the mean). You should be rigorous in your statistical analysis (e.g., there is no such thing as "almost significant"—you set the level of significance required and your data either meet the level of significance or they do not).

RESULTS THAT ARE ALMOST SIGNIFICANT ARE NOT

Sam prepared a manuscript describing her research results. Before performing the research, Sam had set the significance level at 0.05. When Sam conducted the research and performed the statistical analysis, she found that one of the 20 tests was statistically significant at the 0.05 level. Three of the 20 tests achieved significance levels between 0.05 and 0.10. Sam presented the results and went on to discuss them as if all four of these tests confirmed the hypotheses they were designed to test. Sam was surprised when her manuscript was rejected. Reviewer 1 of the manuscript pointed out that Sam should only discuss the single result that reached the level of significance that she had set at 0.05. Reviewer 2 described Sam's results as inconsequential. Reviewer 2 stated that if one test out of 20 was significant, this is only the number of statistically significant results that would be expected by chance. The editor of the journal agreed with both reviewers and rejected Sam's manuscript.

EXPOUND RATIONALLY AND SUCCINCTLY IN YOUR DISCUSSION

The discussion provides an interpretation of findings and their importance within the context of the literature. Typically, the discussion will be organized into three subsections. The first, usually one to several paragraphs in length, will summarize the results of the study, logically developing the argument for how the data address the hypotheses and fulfill the purpose of the study. The second subsection of the discussion will describe how your work fits into the work of others and contributes to the knowledge in the field. This section may be as long as necessary. You should describe the limitations of previous research and how current findings improve or clarify issues. Specify how your research confirms previous findings or provide possible explanations for different results from previous work. The major limitations of your study should be specified. The third subsection is usually only one paragraph long and provides a capsule summary of the study conclusions and any speculations or impact on clinical practice, as well as any future directions for your work. Be cautious not to overstate your conclusions or to speculate excessively.

EDITORS AND REVIEWERS ARE NOT BLIND TO OVERSTATEMENT OF RESULTS AND CONCLUSIONS

Kelly was convinced that his hypothesis was correct. When Kelly prepared a manuscript, he discussed the results as if they had achieved statistical significance. Actually, Kelly had set his level of significant at 0.05. Kelly's "significant" results showed a level between 0.10 and 0.20. If someone read only Kelly's abstract and discussion sections, they would conclude that his results supported his hypotheses and that these results were statistically significant. When Kelly submitted his manuscript to a journal, the editor returned the manuscript to Kelly without sending it out for review. The editor had determined that Kelly's discussion section was not in accord with his results section and rejected the manuscript.

THE REFERENCE SECTION SHOULD BE CHECKED CAREFULLY FOR ACCURACY

You should follow the journal format for the references, and be sure of the accuracy of the reference citations. To check the accuracy of references, go back to the original articles yourself. Also, clarify the journal's policy for citation of in press, submitted, in preparation, and personal communication, since these policies vary among different publications. Some journals require copies of manuscripts that are in press or submitted. Some journals require a letter from the person cited as personal communication.

USE PRIMARY REFERENCES

Lee was in a hurry to prepare a manuscript. Rather than perform a thorough review of the literature, Lee read a recent review and used the citations from the review. One of the reviewers of Lee's manuscript was an author of one of the manuscripts that Lee cited. The reviewer noted that Lee described the reviewer's results incorrectly and listed the reference to the reviewer's study with the wrong volume and page number. Since Lee had also cited the review article, the reviewer looked at the review article and noted that Lee had used the reference citation from the review. The reviewer correctly assumed that Lee had not read the original article. The reviewer summarized these findings in both the confidential comments to the editor and the comments to the author. The reviewer recommended rejection of Lee's manuscript, and the editor agreed. In an attempt to save time, Lee had actually wasted time by not undertaking a thorough review of the literature.

GOOD FIGURES ATTRACT ATTENTION TO YOUR WORK

Tables and figures should contain enough information to stand on their own. You can use figures to sell your manuscript, especially if you invest in color. Some journals have the policy of accepting one or more color figures per article free of charge. A number of journals have a color figure on the cover of each issue. By submitting a color figure with your manuscript, you are in the running for the cover. If your figure is selected as the cover, you should request a copy of the issue. You may want to frame the cover to hang in your office. You should also list the reference in your curriculum vitae as the full reference and add "with cover." A cover figure is an excellent vehicle for marketing your research.

13

•••••••••

HOW TO WRITE REVIEW ARTICLES AND CHAPTERS

•••••••••••••••••

STRATEGICALLY CONSIDER OPPORTUNITIES FOR REVIEW ARTICLES

Throughout the course of your career, you may receive numerous requests to write review articles and chapters. Criteria for promotion and tenure and reviews of grant proposals focus on first-authored, peer-reviewed publications. Review articles and chapters are given relatively little consideration during evaluations for those in tenure-track faculty series, though they may be more important in clinical or master teacher series. In the tenure track, these review articles and chapters will be important only as they are evidence of peer recognition. Therefore, it is important to publish in the most prestigious journals and books in your area. You do not want to spend time writing review articles and chapters because they keep you from having the time to write peer-reviewed manuscripts.

EXPECT AUTHORSHIP FOR WORK YOU AUTHOR, AND ASK AT THE OUTSET

Dr. Smith called graduate student Lynn into the office. Dr. Smith was excited about a request from the editor of a prestigious edited series to write a chapter. Lynn congratulated Dr. Smith on this opportunity. Dr. Smith then suggested that Lynn write the chapter. Lynn was thrilled at this opportunity. However, when Lynn asked about the order of authorship, Dr. Smith said there would only be one author, Dr. Smith. Stunned, Lynn asked if Dr. Smith meant that Lynn would do all the writing and only Dr. Smith would get credit for Lynn's work. When Dr. Smith replied that Lynn was correct, Lynn politely but firmly declined to have anything to do with the chapter. Dr. Smith was unwilling to grant Lynn authorship, and Dr. Smith would either have to write the chapter as sole author or find someone else to be a "silent partner."

WRITE ABOUT WHAT YOU KNOW

Restrict the topic of review articles or chapters to your research area or clinical subspecialty interest. This should be an area that you want recognized as yours. You should already have sufficient knowledge and background in this area to provide a strong foundation for your review of the topic. It should be an area in which you have a reputation or in which you want to develop one.

COAUTHORSHIP CARRIES RESPONSIBILITY

Sandy was a junior faculty member who was flattered when Dr. Jones, a senior professor in her department, invited her to write several chapters for Dr. Jones' edited book. Because this book was considered the "bible" of their field, this would be a big boost to Sandy's career. Sandy was not troubled by Dr. Jones' request to be the senior author on each of the chapters Sandy wrote. Sandy expected that Dr. Jones would contribute substantially to what Sandy produced. Sandy was surprised when Dr. Jones would not read any of the chapters before they were submitted. Sandy had assumed that Dr. Jones would significantly edit each chapter. However, when Sandy proofread the galleys, she discovered that only a few grammatical changes had been made. Sandy was not even sure whether these changes were made by Dr. Jones or a copy editor. Sandy then looked carefully at each of the chapters in the last edition of Dr. Jones' book. Every chapter that was authored by a junior faculty member in Sandy and Dr. Jones' department included Dr. Jones as a coauthor. Sandy wondered why Dr. Jones needed coauthorship on so many chapters in the book Dr. Jones edited.

TURN WORK YOU HAVE ALREADY DONE INTO PUBLICATIONS

Any review articles or chapters you write should be based on your multiple research articles or clinical reports, your recent grant application, or your recent 1-hour talk. It is important to turn reviews for grants or talks into publications, because otherwise, the work you have invested will have been wasted, at least in part. If you have spent the time developing a new conceptual framework to teach or sell your work, you should expand to a broader audience by publication. The timeliness of a review article is determined by a controversy in the area, your ability to uniquely organize the available information, or the need to direct future research efforts.

PEER-REVIEWED RESEARCH ARTICLES SHOULD TAKE PRIORITY

Sam was flattered to be asked to write a chapter by Dr. Smith, a senior faculty member in Sam's department, for a new book Dr. Smith was editing. The topic was slightly outside Sam's research area, but Sam decided the extra research needed to write the chapter was worthwhile. Sam delayed the preparation of several research manuscripts in order to do the library research and writing of the chapter. Sam submitted the chapter on schedule. Six months later, he realized that there had been no communication from Dr. Smith regarding the book. When Sam approached Dr. Smith, Dr. Smith said that the publisher had gone bankrupt and the book would not be published. Sam removed the chapter from his curriculum vitae and tried to make up for lost time on the preparation of research manuscripts. When Sam's curriculum vitae was reviewed for promotion, he was criticized for having too few research publications. Not only wasn't the chapter on Sam's list of publications, all of his work on the chapter had delayed preparation of critical research publications. You can imagine Sam's dismay when two years later, Dr. Smith told him there was good news about the book for which Sam had written the chapter. A new publisher had acquired the rights to the book and wanted to publish the book. Dr. Smith indicated that Sam would have to update the chapter. Sam debated whether he should withdraw the chapter or put in a lot of work to update it. Sam chose a compromise. He added a few new references with a minimal amount of rewriting. Sam also vowed not to agree to write any other chapters unless the publisher was reputable, the editor had an extensive track record of successful books, and the chapter was in Sam's area of expertise.

Know if a Journal Is Interested in Your Review Early in the Process

There are a number of criteria you should use to select an appropriate journal for your review article. If the editor of a journal requests a review article from you, you need to decide if the topic fits the criteria described above and if the journal is one in which you want to publish. If a journal has published multiple research papers or clinical reports on this topic, contact the editor to determine whether there is an interest in your proposed review article. You could also contact the editor of a journal that targets an audience you would like to reach with a review article. Factors to weigh in your selection of a journal include the reputation of the journal and the speed of publication.

Plagiarizing Yourself

Kim was asked by the editor of a journal to prepare a review article on the topic on which Kim had just written a chapter. Kim had a heavy teaching load at the time and submitted essentially the same chapter manuscript to the journal. Kim had forgotten that the copyright to the chapter was held by the publisher of the chapter and not by the author. It is also inappropriate for an author to copy and republish their own work. When the journal editor read Kim's review manuscript, it seemed very familiar. The journal editor checked the chapter, which was referenced in the review manuscript. The journal editor rejected Kim's review article, accusing Kim of plagiarism. The editor told Kim that some journals would sanction Kim for this offense, and would refuse to consider manuscripts from Kim for a period of one or more years.

14

· · · · · · · · · ·

MANUSCRIPT REVIEW

· · · · · · · · · · · · · · · · · · ·

OVERVIEW

Know the Review Mechanism for the Journal You Select

There are various types of review mechanisms. If a journal has a single editor, this editor typically assigns reviewers and makes a decision based on the reviews. When you submit your manuscript to a journal with multiple editors or multiple communicating editors, you select the most appropriate editor, who then may choose the reviewers and make the decision regarding your manuscript. In some cases, the final decision will be made by the editor-in-chief.

Have a Timeline for the Review Process in Mind and Keep Track of It

Lee had just made an important research breakthrough in an area where a number of groups were in intense competition. Lee submitted a manuscript describing the discovery to a prestigious journal and requested an expedited review. The editor agreed to expedite the review of Lee's manuscript. When reading the manuscript to decide on potential reviewers, the editor realized that Lee had just accomplished what the editor's friend had been working on. The editor contacted her friend and told him about Lee's paper. The editor rationalized this breach of ethics based on the strength of the editor's relationship with her friend. The editor slowed the review of Lee's manuscript to allow her friend to complete the work and submit a manuscript to another journal. Once her friend's manuscript had been accepted, the editor accepted Lee's paper for publication. Whereas Lee was concerned that the review had not been expedited, Lee was mostly just relieved to have this breakthrough published. When the editor's friend's manuscript was published a month later, Lee had no way of knowing how the editor had intervened. Perhaps if Lee had pressed the editor to be responsive to the request for an expedited review, the editor may have been forced to process Lee's manuscript more rapidly. Lee should have withdrawn the manuscript and submitted it to another journal when the editor did not provide the agreed-upon expedited review.

Retain a Sense of Control over Your Work during the Review Process

Recognize that you are not completely powerless during the review process. If the evaluation and review are taking longer than you would like, and longer than is usual for that journal, you may withdraw your manuscript. You must do this before submitting it to another journal.

Know the Possible Review Outcomes and Their Implications

There are a number of possible review outcomes. You may be fortunate enough to have your manuscript accepted without any changes. If editors tentatively accept manuscripts, they usually do so contingent upon changes in the manuscript. They may accept a manuscript if the author agrees to shorten it in addition to making other changes, or they may simply request changes. At times editors do not indicate that the manuscript has been tentatively accepted, request extensive changes, and say that the paper will require re-review. Sometimes editors reject the manuscript.

Do Not Let Your Pride Interfere with Publication

Sam made an important discovery and submitted a manuscript to the top journal in the field. Sam was disappointed when the editor requested sev-

eral minor changes to the manuscript, and that Sam shorten the manu-
script to conform with the briefer of the two manuscript formats for the jour-
nal. Rather than make these changes, Sam submitted the full-length man-
uscript to another journal. While the second journal did accept the
manuscript as a full-length article, it was published six months later than
it would have been if Sam had agreed to shorten his manuscript for the orig-
inal journal. In retrospect, Sam wondered if he had made the right deci-
sion, since the second journal was less prestigious than the first one.

Distinguish between Perseverance and Perseveration

If your manuscript has been rejected by three or more journals, you need to decide if you are persevering or perseverating. You should have someone whom you respect read it, someone who will provide an honest opinion of how you should proceed.

Pay Careful Attention to Reviewers' Comments

Chris was a new faculty member, feeling pressure to publish. Chris was
wise enough to ask several senior faculty members to review a manuscript
before submission. All of the faculty members agreed that Chris needed a
lot more data to support the conclusions. Chris dismissed their suggestions
and submitted the manuscript. The first journal rejected the manuscript,
and the two reviewers made points similar to the suggestions made by the
senior faculty members. The second journal returned the manuscript to
Chris without review, saying that the manuscript was inappropriate for
that journal. The third journal had one positive review and one negative
review. The editor of the third journal rejected the manuscript. Chris called
the editor of the third journal and requested a third reviewer, since there
had been one positive review. The editor agreed to contact a third review-
er. The third reviewer had previously reviewed Chris' manuscript for the
first journal. In the comments to the editor, this reviewer was adamant that
the manuscript be rejected, and that the author be told to make the changes
this reviewer had recommended in their initial review for the first journal.
By refusing to listen to the advice of senior faculty members, reviewers, and
editors, Chris was on the way to establishing a very negative reputation.
Senior professors would be unwilling to review future manuscripts. Re-

viewers would remember Chris' lack of substance when reviewing future manuscripts, abstracts for scientific meetings, or grant proposals. Editors would remember Chris' lack of responsiveness when deciding whether Chris' future manuscripts were appropriate for their journal and when considering potential reviewers for manuscripts in Chris' field.

RESPONDING TO THE REVIEWERS' COMMENTS

Read the Editor's Letter as Carefully as the Reviewers' Comments

The editor will send you a letter with the reviewers' comments. Read the editor's letter carefully and make all feasible changes requested. Frequently, the editor may be giving you a message about which of the reviewers' suggestions are most important for you to address.

Make the Easy Changes and Negotiate the Tough Ones

Read the reviewers' comments carefully and make all changes that you can to improve the manuscript, such as following the instructions for authors; spelling, grammar, and nomenclature conventions; stylistic changes; additional information when available; additional data when feasible; and restructuring the discussion to truly reflect the data. There are some changes that you may find difficult to make, such as providing additional data when months of work are required, since these data may be the basis for your next manuscript, or restructuring the paper to reflect a different point of view, which may be consistent with the opinion of the reviewers or the editor, but may not fit with your ideas. Such a change in your position regarding a controversy, model, etc., could have serious consequences to your future work. Therefore, do not mold your opinion just to gain acceptance in a more prestigious journal.

Know When to Move On to the Next Journal on Your List of Three

Sandy submitted a manuscript to a very prestigious and rigorous journal. Reviewer 1 did not agree with Sandy's interpretation of the data and want-

ed Sandy to rewrite the discussion to reflect reviewer 1's point of view. Reviewer 2 felt that Sandy did not have sufficient data and needed to perform a series of experiments that would take about six months to complete. The editor stated that Sandy's revised manuscript would require re-review. In spite of the reputation of the journal, Sandy decided not to include the additional experiments requested by reviewer 2 since they represented the substance of Sandy's next publication. Sandy was also concerned that reviewer 1 had misinterpreted the results. Sandy was unwilling to change the discussion to reflect reviewer 1's opinion. Sandy was also concerned that if the revised manuscript was re-reviewed, it still might not be accepted. Sandy revised the manuscript in accord with the instructions for authors of her second-choice journal and her paper was published within six months.

Your Letter Accompanying Your Revised Manuscript Should Make the Editor's Job Easy

In your cover letter to the editor, you should describe your changes, addressing each point made by each reviewer. Specify the location of the changes in the revised manuscript and comment on any new information. Describe how you changed the original manuscript in accordance with the reviewers' suggestions. If you are unwilling to follow the reviewers' recommendations, give your reasons. Try to accept more reviewers' suggestions than you reject. Thank the editor and reviewers for their comments and tell them how their recommendations have improved the manuscript.

The Editor's Letter May Give You the Editor's Assessment of the Reviewers' Comments

Sam received two reviews of his manuscript. One was relatively positive with a few minor changes recommended. The other reviewer was very negative, requesting major changes, additional experiments, and a significant reduction in the length of the manuscript. In his letter to Sam, the editor suggested that Sam revise the manuscript without performing additional experiments or shortening the manuscript. Sam made the changes that were reasonable without doing additional work or significantly rewriting the manuscript. In his cover letter with the revised manuscript, Sam explained why the additional experiments requested by the second reviewer were unnecessary. The editor accepted Sam's revised manuscript.

How Reviewers Are Selected

Knowledge about the Selection of Reviewers Will Give You an "Edge"

Editors select reviewers from the leaders in the field. They are likely to ask individuals who have given good reviews in the past, who publish in the journal, and who are members of the editorial board. They may contact scientists they have met or heard present at scientific meetings. They may do a keyword literature search to determine who publishes in the area, or they may use the authors of references cited in the manuscript as the basis for an author search. Editors may also use the reviewers suggested by the author after they determine the expertise of the suggested reviewers and their lack of conflict of interest. Conflict of interest would occur if the reviewer and an author were collaborators and had published together recently, or if the reviewer and an author were from the same institution.

Obtain Credit for Your Efforts as a Reviewer

Sandy is a postdoctoral fellow in Dr. Jones' lab. Dr. Jones is an international expert in the field and is often asked to review manuscripts. Even though Dr. Jones is very busy, she always agrees to review manuscripts. She doesn't review them herself, but instead has the students and fellows in her group review them. Dr. Jones never tells the editor that a student or a fellow did the review, and she claims the credit for herself. Sandy is often asked by Dr. Jones to review manuscripts. Sandy enjoys the review process and Dr. Jones praises Sandy's reviews. Sandy is curious about each manuscript he reviews and always reads the paper if it is published. The most recent paper Sandy reviewed is published and Dr. Jones was asked to prepare a commentary to appear with the article. Suddenly, it dawns on Sandy that Dr. Jones used him to write the review that became the basis for the commentary. Not only did Dr. Jones get credit for Sandy's work in doing the review, but Dr. Jones also got credit for the publication of the commentary based on Sandy's review. The next time Dr. Jones asked Sandy to review a manuscript, Sandy stated that he would do the review only if Dr. Jones told the editor that Sandy was the reviewer and if Sandy could communicate the review directly to the editor. Faced with Sandy's ultimatum, Dr. Jones agreed.

HOW TO REVIEW A MANUSCRIPT

Accept the Responsibilities of Review Conscientiously

When you are asked to review a manuscript, accept if you can meet the deadline, if the topic is within your area of expertise, and if there is no conflict of interest. You should reserve your comments for substantive issues. You can annotate the manuscript to correct spelling, grammar, and style. You should return your review in a timely fashion. Remember how frustrating it can be when your manuscripts are delayed due to slow or unwilling reviewers. Holding up a competitor's paper is unacceptable. Review a competitor's paper the same day you receive it. Information in the manuscript is confidential. You may not cite it until the paper is published or you receive permission from the author. As a reviewer, you may contact the editor to determine the status of the manuscript, though many journals will let you know the outcome of the review and may even give you a copy of the other reviewers' comments. If the manuscript was accepted, you may ask the author whether you may cite their manuscript. Editors remember good reviewers when the reviewers submit manuscripts of their own or when it is time to add someone to the editorial board. Editors are also likely to serve as reviewers for abstracts for scientific meetings. Serving as a reviewer or editor allows you to help shape your field—publishing good work and keeping the bad science out of the literature.

If You Write, Then You Must Be Willing to Review

Kim is aware that as an assistant professor, peer-reviewed publications are important to her promotion and tenure. Kim submits manuscripts at the rate of one per month. Kim's publication rate has impressed the senior faculty in the department. Kim often asks editors to expedite the review of her manuscripts. However, Kim never agrees to review a manuscript. Finally, one editor, who had published several of Kim's manuscripts, was frustrated by Kim's fifth refusal to review a manuscript for the editor's journal. The editor spoke to Kim, and reminded her that reviewers were an important part of the journal infrastructure. If everyone behaved like Kim, there would be no peer review of manuscripts.

How Editors Are Selected

Editors Are Selected by a Variety of Mechanisms

Editors for some journals are elected. They are nominated by their colleagues and elected by the membership of the organization that supports the journal. Other editors are selected from the editorial board of a journal on the basis of their abilities as reviewers and their commitment to the journal. Some journals advertise before the editorship is vacated. If you are interested in such a position, ask your senior colleagues if they think you would be appropriate and whether they would be willing to write a letter of nomination on your behalf.

Becoming an Editor Requires Preparation

Lee is a very ambitious assistant professor. One of Lee's goals is to become the editor of a major journal. In order to achieve this goal, Lee decides the first step should be membership on the editorial board of a major journal. In order to take this first step, Lee decides to write to the editors-in-chief of several major journals, requesting to be placed on the editorial board. In support of his request, Lee sends a cover letter and his curriculum vitae. While Lee has a number of publications, he has only reviewed one or two papers each for about five journals. Some of the journals Lee writes to are ones for which he has never served as a reviewer. Most of the editors simply discard Lee's letter and vitae. One takes the time to write to Lee and explain that members of the editorial board are selected based on extensive, skilled reviewing for the journal for a number of years, as well as an international reputation in the field. In addition, editorial boards need to have a geographic distribution and a varied composition of members from each of the areas covered by the journal.

chapter

15
·········

ETHICAL BEHAVIOR

····················

DEVELOP FUNDAMENTAL PRINCIPLES WITH WHICH YOU WILL CONSIDER ETHICAL DILEMMAS

Becoming an academician includes formal and informal training on ethical behavior. All NIH-funded training programs require that all trainees participate in a formal course on ethical behavior. These are typically presented over a two-day period. In addition to coursework, you should also observe those around you. Mentors provide training in ethical behavior through both exemplary behavior and discussions of dilemmas they have faced or discussed with others. Many academic settings provide ombudspersons to deal with ethical concerns and interpersonal conflicts in a confidential manner. You must recognize that for most ethical dilemmas there is no one single correct solution. Therefore, it is important for you to develop your own fundamental concepts and principles with which you will be able to consider dilemmas as they arise. The following examples are written to stimulate you to consider some of the ethical issues that you may face during your academic career. Your mentors and your institution's ombudsperson can provide advice to you when you face such dilemmas.

ETHICAL SITUATION 1: YOU HAVE A RIGHT TO INFORMATION CONCERNING YOUR OWN CAREER

Kim was enjoying an academic career and felt successful during the first year. At the end of each year, the department chair and the associate dean for academic affairs invited each faculty member to discuss the past year and receive feedback. Kim thought that the meeting with the department chair went very well, so well that Kim did not ask to see the report that the department chair forwarded to the associate dean. Kim's meeting with the associate dean did not go well at all. It was as if Kim were two different people. Fortunately, she went to a senior faculty member of the department to discuss this discrepancy. The senior faculty member asked whether Kim had read the department chair's report on Kim's performance. When Kim replied that she had not, the senior faculty member suggested that perhaps Kim should make another appointment to meet with the associate dean to discuss the department chair's report and the discrepancy between Kim's meetings with the chair and with the associate dean. During this second meeting with the associate dean, Kim read the department chair's report and found that it was as negative as Kim's meeting with the department chair had been positive. The associate dean offered to meet with the department chair to discuss the discrepancy. The dean suggested that the chair had preferred another candidate for Kim's position, and that perhaps the chair wanted to encourage Kim to leave so that the other candidate could be hired.

It is very important for you to review the information in your files that are open and available to you. You may find discrepancies with your own impression of your progress, and these insights will help you improve. There also may be outright errors that you need to correct. If you feel that there is a discrepancy between the verbal and written comments about you, then you need to try to clarify it. There have been examples of written documentation of performance being used both to encourage people to leave and to force them to stay (by limiting their opportunities to leave).

ETHICAL SITUATION 2: DO NOT AGREE TO CONDITIONS OF EMPLOYMENT WITH WHICH YOU CAN NOT COMPLY

When Lee joined Dr. Smith's lab group as a postdoctoral fellow, he agreed that the project he worked on would remain in Dr. Smith's lab when he left for a faculty job. This was standard practice in Dr. Smith's lab and all trainees made the same agreement before they began to work with Dr. Smith. Two years later, Sam, a graduate student in Dr. Smith's lab, finds Lee furtively copying his lab notebooks on the eve of his departure for his first

academic position. When Sam asks Lee the purpose of copying the lab note-
books, he replies that he plans to continue his project as a faculty member
at another institution. When Sam reminds Lee that they had all agreed not
to take their research with them when they left, Lee replied that he had
changed his mind and any such agreement with Dr. Smith was bogus. The
first thing the next morning, Sam speaks to Dr. Smith about Lee's behavior.

It is distressing when a mentor does not help a trainee move to the next level, including limitation of project movement by postdoctoral fellows. Most areas are broad enough to permit the fellow some opportunity to continue their momentum. On the other hand, if the trainee has entered into an agreement, they should honor it. While there is probably no legal foundation for the agreement, they will need letters of reference from their mentor. The trainee who is party to such an agreement should be considering, and even developing, alternative projects in the course of their training.

ETHICAL SITUATION 3: SEXUAL HARASSMENT AND OTHER FORMS OF POWER ABUSE SHOULD NOT BE TOLERATED

Stacey enters the restroom to find a very upset student, Lee. When Stacey
asks Lee what the problem is, Lee replies that Dr. Jones (a very senior fac-
ulty member in Stacey's department) has demanded sexual favors in re-
turn for awarding the grade Lee has earned in Dr. Jones' class. When Lee
told Dr. Jones to bug off and threatened to report Dr. Jones to the depart-
ment chair, Dr. Jones replied that Dr. Jones would see that Lee would nev-
er be able to pursue graduate studies. Stacey suggests that they both meet
with the department chair to discuss Dr. Jones, but Lee refuses due to con-
cern about a future career. Stacey went to the chair of the University Com-
mittee on Sexual Harassment. The committee chair assured Stacey that
their inquiry would protect Lee's identity. When Stacey next spoke to Lee,
Lee was willing to meet with the Committee on Sexual Harassment. It was
revealed that Dr. Jones had solicited sexual favors from a number of stu-
dents in exchange for grades. Several students agreed to go to the Commit-
tee on Sexual Harassment. As a result of this inquiry, Dr. Jones retired im-
mediately and was unable to interfere with the future careers of the
students who testified to the Committee.

Sexual harassment, which may occur among any gender combination of teacher/mentor and student/trainee, is unconscionable and illegal. If the student/trainee is under administrative control in any way by the teacher/ mentor, and the institution has not set up a specific mechanism (e.g., ombudsperson) to deal with this problem or other examples of power abuse, then the student/trainee needs to identify someone with sufficient sensitivity and security to take on this issue. In the absence of such an individual, talk to trusted friends and colleagues outside of the institution for guidance. These relationships are not only disconcerting, they are potentially dangerous, mentally and physically, for the student/trainee.

ETHICAL SITUATION 4: RESULTS OF INVESTIGATIONS SHOULD BE PUBLISHED

Chris participated in a research project as a second-year graduate student. A postdoctoral fellow took the lead on the project and was first author on the abstract; Chris was the second author. The postdoctoral fellow left for a job in industry. Chris asked the postdoctoral fellow repeatedly over a two-year period about writing up the manuscript. The former postdoctoral fellow claimed no interest in the paper, because it was irrelevant to the job in industry. Chris went to senior investigator in the lab and offered to write the manuscript as long as Chris would be the first author.

It is important for data to be published. The resources involved in obtaining the results are enormous. As individuals change jobs, they may lose interest in publication. Someone else can step in to facilitate writing up the project but, if possible, the original individual should be kept involved, though probably not as first author.

ETHICAL SITUATION 5: AVOID OVERINTERPRETATION OF DATA

Sandy met with his mentor to present some new data. Sandy's mentor was excited about Sandy's results and suggested that Sandy reanalyze the data to emphasize some small differences between the two groups. Sandy was very concerned by this request and met with the department chair to discuss these issues. The department chair praised Sandy and suggested that they meet with Sandy's mentor to prevent overinterpretation of Sandy's data.

Such situations may represent overinterpretation of results. On the other hand, the mentor may have insight into the results and methodologies that permits an interpretation not understood by the mentee. In such situations, it is critical for the mentee to get a senior person involved and ask them to review the results and provide their interpretation. It is best if the mentor is also engaged in this process.

ETHICAL SITUATION 6: HONESTY IS AN ABSOLUTE REQUIREMENT IN SCIENCE AND MEDICINE

Chris and Stacey are postdoctoral fellows in the same laboratory. Chris' research has been progressing very well. Chris already has two papers in top-level journals. Stacey's project is moving a lot slower. Chris often reviews Stacey's data in order to help Stacey. Chris is surprised when Stacey presents data at a lab meeting that is the exact opposite of the data Stacey has been getting week after week and showing to Chris. The data Stacey presents send their mentor into ecstasy. These data are just the thing to support a major aspect of the mentor's favorite theory. When Chris asks Stacey what changed to get the new data, Stacey brushes Chris off. Chris is concerned that Stacey made up the results to please their mentor. Chris meets with the mentor to discuss these issues. When the mentor confronts Stacey, Stacey confesses to making up the data and is dismissed from the laboratory.

Fabrication of data is absolutely unacceptable. In science and medicine, we must be as honest as humanly possible, because so much of what we do requires people to trust us. There should, however, be due process before summary dismissal.

16

LEADERSHIP

YOU ARE A LEADER

From your selection of an academic career, you are clearly a leader. You have already assumed a number of leadership roles: as a teacher, as a researcher, as a professional, as a member of professional organizations. Your continued participation in the academic infrastructure is essential to the future of your discipline and the future careers of your mentees.

WHY DOES SOMEONE WANT TO BECOME A LEADER?

Leaders may have a new vision for the group and want to implement it. They may want to engage in creative problem-solving. They may want to have a larger impact as a mentor and help to develop the careers of others.

LEADERS MUST HAVE APPROPRIATE PREPARATION TO ESTABLISH LEADERSHIP SKILLS

Kim is a full professor and a successful researcher. Kim focused on her research to the exclusion of other aspects of her career, steadfastly refusing to serve on committees, take on extra teaching responsibilities, or pursue additional service commitments. Looking for a new challenge, Kim decides to become a department chair. Kim meets with her current department chair to discuss her new goal. Kim's department chair is caught totally off guard by her plans. The chair tries to explain that one becomes a department chair by demonstrating leadership in a variety of areas, including research, administration, teaching, and service. Since Kim had always been very frustrating to the department chair by refusing to participate in the departmental infrastructure, the chair has a difficult time reconciling her past behavior with her new goal. The department chair suggests that Kim participate more fully in the department in order to test the reality of her new goal.

EVEN IN GROUPS OF TWO INDIVIDUALS, ONE WILL BE A LEADER

Any productive group of two of more individuals will involve leadership to keep the group moving toward a common goal. With two individuals, the leadership may alternate between the duo, depending on the nature of the issue being addressed. With three or more people, there may be layers of leadership so that, for example, the more senior trainee, mentored by a faculty member, may be showing leadership in working with the more junior trainees in the group.

WHAT ARE THE OPPORTUNITIES FOR LEADERSHIP IN RESEARCH?

The first step, as a faculty member, is the establishment of your own research group and your independence as an investigator. Another leadership role is serving as the principal investigator of a research grant. You may progress to being the principal investigator of a program project or center grant. Your research role may eventually expand to being the director of a research institute.

THE MOST JUNIOR MEMBER OF A GROUP MAY FIND LEADERSHIP ROLES

Chris had always been a leader but was dismayed when participating in research for the first time as part of a new training program. Chris was struggling to master all the skills required for research and was feeling very insecure. What could Chris do to show leadership while being the most junior person in terms of scientific experience? Chris could take on some teaching roles by asking thoughtful questions, doing literature searches, and volunteering to do presentations or teach methods to other lab members. Chris could organize social activities for the lab group: potluck lunches, birthday cakes, intramural sports teams, or group participation in a fundraising activity. Chris could volunteer to be responsible for a critical task for lab performance, such as ordering supplies, equipment maintenance, or inventory.

EDITORIAL AND REVIEW ACTIVITIES AFFORD OPPORTUNITIES FOR LEADERSHIP

Another route for leadership in research is through editorial activities. Serving as a reviewer is a beginning leadership opportunity. You may progress to membership on an editorial board or function as a communicating editor, or editor. The importance of this type of leadership is that you are shaping the field by recommending rejection of manuscripts that should not be published and working to improve acceptable manuscripts to become building blocks for further development of the discipline. As an editor, you also have the opportunity to publish single issues devoted to a specific topic, discipline, or methodology. Editors can change the direction of the journal to reflect new areas of interest within a field. Editors can also develop different formats. For example, there is currently a great deal of interest in commentaries on the major articles in some of the more prestigious journals.

TEACHERS ARE LEADERS AND WRITERS ARE TEACHERS

Leadership activities in education include serving as a course organizer and lecturer. You might become more involved in curriculum development, establishing new courses or totally rewriting the curriculum. If you decide to expand your leadership role to textbook publishing, you could serve as author of one or more textbook chapters, edit a multiauthor textbook, or author a textbook. If you are interested in writing a book, you will need to have a very thorough outline developed before you begin to contact publishers. You should contact those publishers who produce books in your area of interest. Publishers will be more interested in your ideas if you have evidence of previous authorship or editorship.

ADMINISTRATION OFFERS MANY OPPORTUNITIES FOR LEADERSHIP

Involvement in administrative leadership could begin with committee membership. Your role could expand to a committee chair. Further administrative leadership would be shown as division head or department chair. You might consider moving up to assistant or associate dean, dean, or chancellor. Each of these administrative roles requires previous experience demonstrating administrative skills. Administrative roles at the department chair level and above may require a move to a different institution. One way for institutions to keep from becoming stagnant is to recruit administrative leaders from outside the institution. They are able to do this if the institution is willing to provide enough institutional resources to support the recruitment of a leader with vision. Without such resources, it will be difficult to attract someone from the outside. Typically, it costs less to promote an internal candidate into such an administrative role.

PROFESSIONAL ORGANIZATIONS ARE VENUES FOR LEADERSHIP

You may choose to seek leadership roles in professional organizations at the local, state, regional, national, or international levels. You could serve as a committee member, committee chair, or officer. By being a member of a professional organization, you are already supporting the infrastructure of the discipline, which may include meetings, educational material, journals, training programs, or policy. Through committee work or as an officer, you can have a decision-making impact, moving the organization in a particular direction.

THE CHOICE OF WHICH LEADERSHIP ROLES YOU PURSUE IS YOURS

You should engage in activities to which you are committed. Leadership activities are important for promotion and tenure. They also afford you the opportunity to have a lasting impact on your field. But you must devote yourself to each activity with your fullest attention or you will not be credited with having leadership potential.

chapter

17

• • • • • • • • • •

PREPARING A CURRICULUM VITAE

• • • • • • • • • • • • • • • • •

YOUR CURRICULUM VITAE IS A REFLECTION OF YOUR PROFESSIONAL CAREER

Your first contact with a potential mentor or department chair is often through your curriculum vitae. Your curriculum vitae provides the information for the NIH biosketch included in every NIH grant application, your materials for promotion and tenure, your portion of the department Web site, and any other application that requires historical information about your professional career. You will also be asked to provide a curriculum vitae when you are appointed to committees of professional organizations, when you are invited to give a seminar, or when you teach a course for continuing education credit. In addition to providing an opportunity to make a good first impression, your curriculum vitae provides for systematic storage of information that you may need to respond to any request for professional biographical information, frequently on short notice.

YOUR CURRICULUM VITAE IS NOT A BUSINESS RESUME

This document should have the heading "Curriculum Vitae." Technically, curriculum vitae refers only to your professional history, with the bibliography being a separate document. However, by common practice, both are usually embodied in a single curriculum vitae document. A curriculum vitae is not a business resume. While business resumes are organized from current to oldest, curriculum vitae should be organized from oldest to most recent. For example, you should list your employment from the first job through your curent postiton.

Include Your Name, Titles, "Coordinates," and as Much Personal Information as Is Comfortable for You

The first piece of information is your name, including your degrees and your current position. Be sure to include all of your professional titles, including membership in organizational research units, centers, or institutes outside of your primary departmental affiliation. You should then provide your current address, phone, fax, and e-mail. You may wish to include your social security number, which may be needed if you are to receive compensation or reimbursement. Other optional information includes your birthplace and birth date. Some people include information about their family including their marital status, length of marriage, significant other's name, significant other's occupation, and children's names and ages. Information about your household is useful during recruitment to assist your significant other in finding employment, to facilitate locating schools for your children, and to suggest potential neighborhoods for housing.

Your Educational Information Should Provide an Accurate and Complete Review of Your Professional Preparation

Information regarding your education and training should include the time period, discipline, institution, city, state, degree, and date of degree. Description of your professional experience should include the time period, your title, the department, institution, city, and state. If you took time off or had breaks in your education or professional experience, consider explaining these briefly, because the speculation about these "holes" in your career may be damaging to you as a trainee or job candidate.

Your Honors Represent Important Distinguishing Characteristics

Under "Honors," you should include election to any honor societies, graduation with honors, fellowships, scholarships, awards, and named or memorial lectures you have presented. You should include the date that the honor was received.

COMMITTEE MEMBERSHIPS SHOW GOOD CITIZENSHIP AND LEADERSHIP

You should specify your membership or chairmanship of any committees outside of your employing institution. These committees may include professional organizations, volunteer organizations, and grant review panels. Listings should include the time period, your role, the name of the committee, the organization, and any other pertinent information.

ORGANIZE YOUR ACTIVITIES BY INSTITUTION

For each institution where you have had an appointment, indicate your teaching, committee, and professional or clinical activities. Each of these entries should include the time period you were involved, your role, and the name of the activity. Under teaching, you should include lectures you presented, courses you organized, curricula you developed, and training programs in which you were involved. Under committees, list the duration of your involvement, the committees in which you were involved, and your role. Under professional or clinical activities, you should detail the duration, the nature of these activities, and your role.

INCLUDE ANY PROFESSIONAL LICENSES OR CERTIFICATES

Indicate your professional certifications, the dates they were obtained, and any identification numbers. You should also include your professional licenses, the dates they were obtained and the identification numbers. Some jobs you apply for may require certification or licensure.

LIST YOUR GRANTS AND STUDENTS/TRAINEES

List all of your successful and currently pending grant proposals, providing duration, title, total award amount, and your role. If there are grants that you were awarded and could not accept, for example, because of duplication of awards, then you should include those and list them as awarded but declined, with the reason briefly stated. List all of your trainees, their training duration, their research projects, and their dates of degrees and training institutions. You may also wish to include awards your mentees received and their current positions and institutions.

SEPARATE YOUR PEER-REVIEWED PUBLICATIONS FROM THE OTHER CATEGORIES

In your bibliography, you should include several categories of publications. Each of the following should have a separate section: peer-reviewed publications; reviews and chapters; and other publications. Published abstracts should follow. The last section should include your oral presentations, with the topic, type of talk, department, institution, city, state or country, and date. If the oral presentation was at a professional meeting, you should include the topic, meeting, city, state or country, and date.

YOUR CURRICULUM VITAE IS A REPOSITORY OF VALUABLE INFORMATION, SO KEEP IT UP-TO-DATE

The purpose of a well-organized and current curriculum vitae is to keep records to facilitate your grant proposals and documentation to support your promotion and tenure. As your career progresses, it will become increasingly difficult to recover lost information. You will be wise to keep your CV updated on a weekly basis, as new information becomes available. Do not put this off or information will be lost. Keeping this document current will facilitate your ability to respond to requests for career information stored in your curriculum vitae. It is disconcerting to be given someone's curriculum vitae, for example, to support their job candidacy and realize that it is one or more years out of date. Such delinquency raises serious concerns about your organizational skills.

18

••••••••••

SUMMARY

Gauging Success

•••••••••••••••••

YOU WILL HAVE MORE THAN ONE MENTOR

In order to succeed in academics, you need to select and utilize a number of mentors. Different mentors are required for different times and different aspects of your career. New mentors are needed as you grow and acquire new responsibilities. Remember also, that mentors are not always senior to you in the hierarchy. You will find that your students and trainees can have tremendous insight into your strengths and weaknesses and can give you outstanding advice.

YOUR MENTORS WILL BE CONCERNED ABOUT YOUR PROFESSIONAL DEVELOPMENT

There are a number of qualities that are important in your mentor. These individuals should have demonstrated strengths in your area of interest. They should be willing to invest the time and effort to provide effective mentoring. They should be concerned with your personal development. They should have a track record in career development and a commitment to further your career. Perhaps the most important characteristic of your mentor is personal integrity.

YOUR MENTOR WILL HELP YOU ACHIEVE INDEPENDENCE

Your scientific mentor should help you to develop into an independent investigator. They should encourage excellence and scientific integrity, while remaining sensitive to your needs. They should teach principles, judgment, skills, and perspective. They should encourage you to accept your strengths and weaknesses and show you how to adapt to organizational realities. They should provide opportunities to develop independence by having you give scientific presentations, write manuscripts, and prepare grant proposals.

DEVELOPING RESEARCH SKILLS WILL TAKE TIME AND MAY BE FRUSTRATING

While research is important for an academic career, it is often difficult for graduate students and clinical trainees to make the transition into a research career. When you are frustrated by the slow progress of your research, you need to remember how many years and how much effort you expended in acquiring other skills. You will need to invest a similar amount of time and effort in acquiring laboratory technical skills, as well as the skills involved in sharing your work (preparation of abstracts, presentations, manuscripts, and grants). Just as you acquired other skills by accepting new challenges, you need to watch other people in the lab perform techniques, seek to learn new methods, and look for opportunities to discuss your research. Instead of kicking yourself for not having a major grant during your first year doing research, consider how far you have come from when you began your research, what you have learned, and what progress you made. Break each major research goal down into subgoals and celebrate reaching each of the subgoals.

YOU ARE A MENTOR

Teaching is fundamental to any academic career, and teachers are mentors. Therefore, as you teach students and trainees, you are their mentor. Remember the characteristics you value in a mentor and demonstrate these characteristics for your mentees.

YOU ARE IN CONTROL OF YOUR CAREER

Taking control of your career empowers you to succeed in an academic environment. Your career is what you make of it. Look at your publications and grants as the currency of your career. Academics provides a unique opportunity to sell your ideas and to achieve a level of independence that is rare in other fields. Good luck!

INDEX

A

Abstracts
  research paper preparation, 112
  scientific meeting presentations, 73–80
    clarity, 78–80
    components, 75–78
      authors, 75–76
      conclusions, 77
      introduction, 76
      methods, 77
      purpose, 76–77
      recommendations, 78
      results, 77
      speculation, 78
      title, 75
    electronic submissions, 79–80
    focus, 74, 83
    goals, 74
    overview, 73
    preparation time, 78–80
    purpose, 74–75
    writing style, 74, 80
Administration
  grant review, 59–60
  leadership opportunities, 138
Agency for Health Care Policy and Research,
  web site, 33
Authorship, *see* Publication; *specific writings*

C

Career development
  academic position selection
    career path protection, 25, 27–28
    contracts, 26–27, 100, 130–131
    dual career couples, 5–6
    evaluation criteria, 26, 29
    interviews, 27–29
    job search, 23–24
    lateral moves, 23
  curriculum vitae preparation, 139–142
    activities, 141
    committee membership, 141
    description, 139–140
    educational information, 140
    grants, 141–142
    honors, 140–141
    organization, 139, 142
    peer-reviewed publications, 142
    personal information, 140
    professional licenses and certificates, 141
    trainees, 141–142
  ethical behavior
    career information, 129–130
    confidentiality, 113
    employment conditions, 130–131
    honesty, 133
    power abuse, 131–132